Core Mathematics

Dedicated to my maternal uncle and aunt

Core Mathematics

George N. Frempong

Printed by CreateSpace

www.CreateSpace.com/TITLEID

Available from Amazon.com and other retail outlets

Contents

Preface

This book is second of two volumes written for students following the Core Curriculum, and is ideal for students preparing for Cambridge IGCSE, GCE O Level Mathematics and the West African School Certificate Examination.

This book has been developed from materials used in teaching mathematics at Accra High School for twenty eight years.

I have tried to present the concepts in forms that are easy to understand. Each of the sections offers a step-by-step explanation of the concepts together with many worked examples. The Try this exercises give you the opportunity to practise after every example. The many exercises allow you to have as much practice as possible. Every chapter ends with a test. The tests have been provided to help you track your progress and also help you retain mastery of the topics.

It will not be possible to look back and recall the origins of most of the materials in this book. I am very grateful to all whose work I have benefited from.

1

Trigonometry

Trigonometry is a branch of mathematics that studies triangles and the relationship between their sides and the angles between sides. Trigonometry can be used to calculate the lengths of sides and sizes of angles in triangles.

Trigonometric Ratios

A ratio of the lengths of two sides of a right-angled triangle is called a trigonometric ratio. Trigonometric ratios are related to the acute angles of a right-angled triangle. The three basic trigonometric ratios are sine, cosine and tangent, abbreviated sin, cos and tan respectively.

The tangent ratio

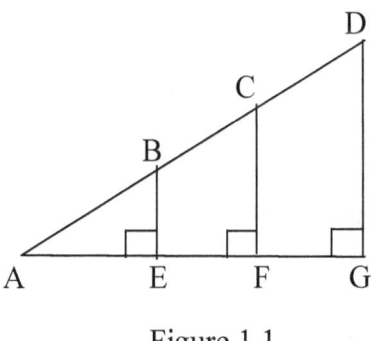

Figure 1.1

Figure 1.1 shows three overlapping right-angled triangles $\triangle ABE$, $\triangle ACF$ and $\triangle ADG$ that share a common angle, $\angle A$. All three triangles are similar to one another. So $\dfrac{BE}{AE} = \dfrac{CF}{AF} = \dfrac{DG}{AG}$

In any right-angled triangle, the ratio of the length of the side opposite an acute angle to the length of the side adjacent to that angle is a constant called the tangent of the angle. Notice that the ratio of the sides of a right-angled triangle is determined solely by the size of the acute angles but not by the size of the triangle.

Definition of Tangent

The sides of a right-angled triangle have names relative to the acute angles, as shown in Figure 1.2

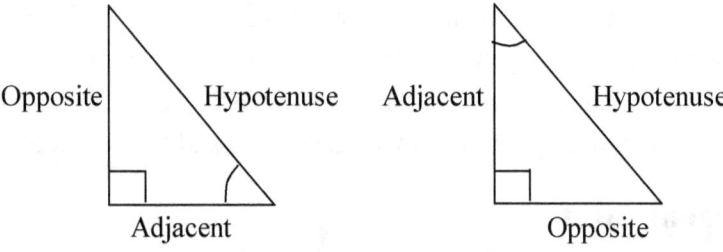

Figure 1.2

The side opposite the right angle is called the hypotenuse. The hypotenuse is the longest side of a right-angled triangle. The side opposite a given angle is called the opposite side and the side next to the angle is called the adjacent side. Note that the opposite and adjacent sides relate to the acute angle under consideration.

We define the tangent of an angle θ as

$$tan\ \theta = \frac{length\ of\ side\ opposite\ angle\ \theta}{length\ of\ side\ adjacent\ to\ angle\ \theta}$$

Try this 1

The triangles shown below are right-angled triangles. Read off the tangent of each acute angle in these triangles without simplifying

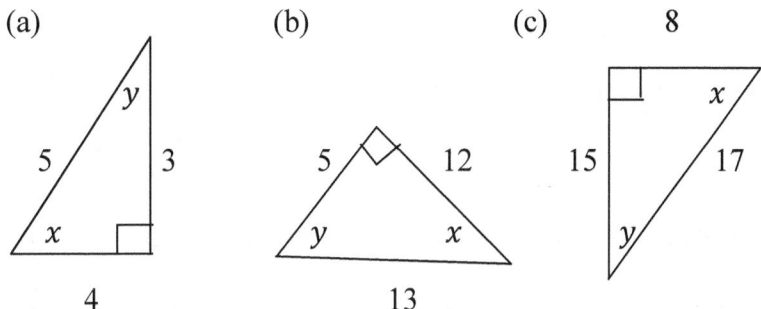

Finding Tangents of Angles

You can use your scientific calculator to obtain tangents of angles and angles whose tangents are given. To find the tangent of an angle, your calculator may expect you to key the TAN button before you enter the angle. Some calculators will require that you enter the angle first before you key the TAN button. Make sure you know how your calculator works

Example

Find Tan 75°, using your calculator

Enter: [TAN] [75] =

You should be given the answer 3.732

Try this 2

Find Tan 58°, using your calculator

The example below illustrates how you can use your calculator to find an angle when the tangent is given.

Example

The tangent of an angle is 1.881 find the size of the angle.

Enter: [SHIFT] [TAN] [1.881] =

You should be given the answer 62^0

Try this 3

The tangent of an angle is 0.7813 find the size of the angle

Remember that some calculators will require you to enter the number first.

Exercise 1.1(a)

1. Find, correct to 4 significant figures, the tangent of the following angles:

(a) 50^0 (b) 79^0 (c) 35^0 (d) 68^0

(e) 73.5^0 (f) 40.8^0 (g) 75.28^0 (h) 85.02^0

2. Find correct to 1 decimal place the size of an angle if the tangent of the angle is:

(a) 0.7265   (b) 2.356   (c) 11.43   (d) 1.091

(e) 6.071   (f) 1.317   (g) 6.520   (h) 9.845

Solving Right-Angled Triangles using Tangents

Given the length of one side of a right-angled triangle and the size of one of the acute angles, you can find the length of the other sides. Also, you can find one of the acute angles in a right-angled triangle when you know the length of two sides of the triangle.

Finding the length of a side of a right-angled triangle

Examples

In the diagram below, PQR is a right-angled triangle, $\angle QPR = 90°$, $\angle PQR = 36°$ and PQ = 5 cm, calculate the length of PR, correct to 1 decimal place.

First, draw a diagram that represents the given information

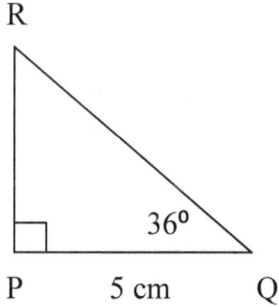

R

36⁰

P   5 cm   Q

From the diagram, you can see that PR is the opposite side and PQ is the adjacent side to angle 36⁰.

$$\frac{PR}{5} = \tan 36°$$

$$PR = 5 \tan 36°$$

$$= 5 \times 0.7265$$

$$= 3.6 \text{ cm}$$

Try this 4

Sketch the right-angled triangle PQR with $\angle PQR = 90°$ $\angle PRQ = 48°$ and QR = 15 m. Calculate the length of PQ.

In the diagram below,   ABC is   a right   triangle.   AC = 12 cm, $\angle BAC = 90°$ and $\angle ABC = 65°$, calculate the length of AB correct to 1 decimal place.

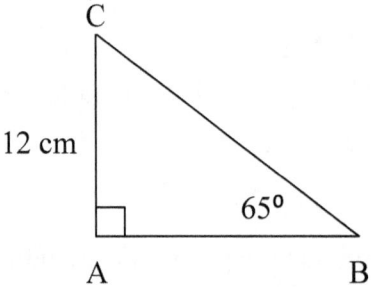

From the diagram, you can see that AC is the opposite side and AB is the adjacent side to angle 65⁰.

$$\tan 65° = \frac{12}{AB}$$

$$AB \times tan65° = 12$$

$$AB = \frac{12}{\tan 65°}$$

$$= 5.6 \text{ m}$$

This result can also be obtained in the following way:

First find $\angle BCA$

$$\angle BCA = 90 - 65 = 25°$$

Now, the side AB is the opposite side and AC is the adjacent side to the 25^0 angle. Therefore,

$$\frac{AB}{12} = \tan 25°$$

$$AB = 12 \tan 25°$$

$$= 5.6 \text{ m}$$

Try this 5

Sketch the right-angled triangle ABC with $\angle ABC = 90°$, $\angle BCA = 56°$ and AB = 20 cm. Calculate the length of BC.

Exercise 1.1(b)

The triangles shown below are right triangles. Calculate the length of the side marked x on each of the following diagrams. Give your answer to one decimal place.

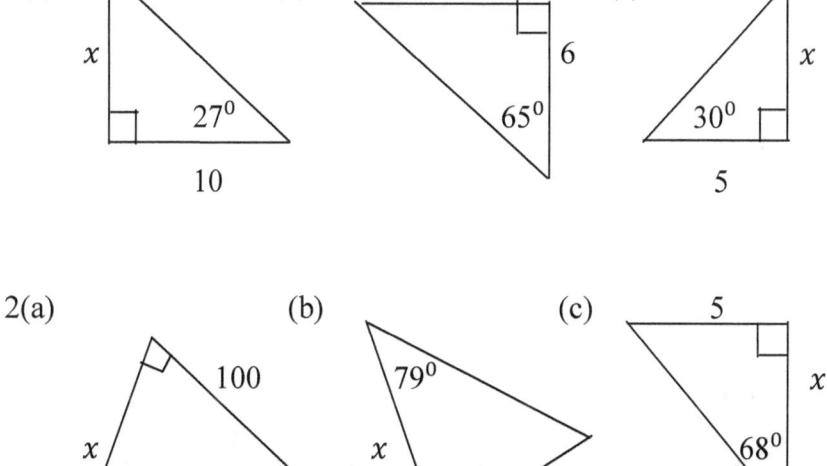

1(a)   (b)   (c)

2(a)   (b)   (c)

Finding an Angle of a Right-Angled Triangle

Example

In the diagram below, ABC is a right-angled triangle. $\angle BAC = 90°$, $\angle ABC = x°$, AB = 4 cm and AC = 5 cm find the size of angle x. Give your answer correct to 1 decimal place

First, draw a diagram that represents the given information

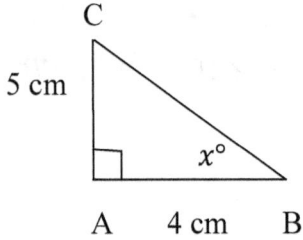

You can see from the diagram that AC is the opposite side and AB is the adjacent side to angle x.

$$\tan x° = \frac{5}{4}$$

$$= 1.25$$

$$x = 51.3°$$

Try this 6

Sketch a right-angled triangle with $\angle PQR = 90°$, $\angle QRP = x°$, PQ = 15 m and QR = 20 m. Find the size of angle x, correct to 1 decimal place.

Exercise 1.1(c)

The triangles below are right triangles. Calculate to one decimal place the size of the angle marked x in each of the following diagrams

1(a)   (b)   100   (c)

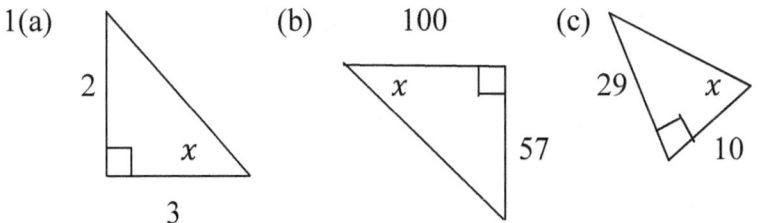

2(a)   (b)   15   (c)

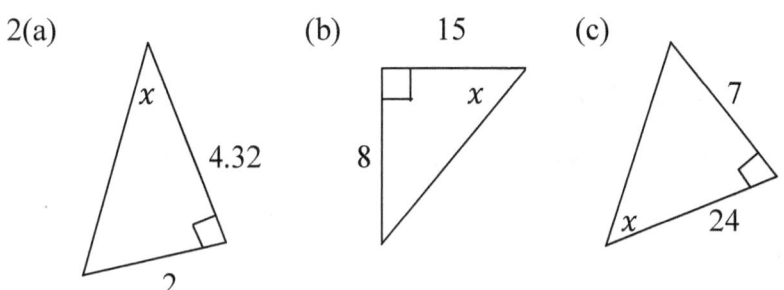

Example

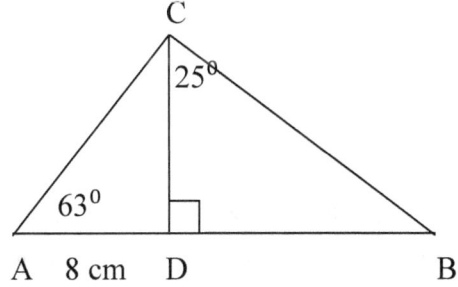

In the diagram above $\angle CAD = 63°$, $\angle BCD = 25°$, $\angle ADC = 90°$ and AD = 8 cm. Find the length of BD. Give your answer to one decimal place

First, use triangle ADC to find the length of CD. Note that triangle ABC is not a right-angled triangle.

In triangle ADC, CD is the side opposite angle 63^0.

$$\frac{CD}{AD} = \tan 63°$$

$$CD = AD \tan 63°$$

$$= 8 \times 1.963$$

$$= 15.704$$

Next, use triangle CDB to find BD.

In triangle CDB, BD is the side opposite angle 25^0

$$\frac{BD}{CD} = \tan 25°$$

$$BD = CD \tan 25°$$

$$BD = 15.704 \times 0.4663$$

$$= 7.3 \text{ cm}$$

Exercise 1.1(d)

Calculate the length of the side or size of the angle marked x in each of the following diagrams. Give your answer to 1 decimal place.

1.

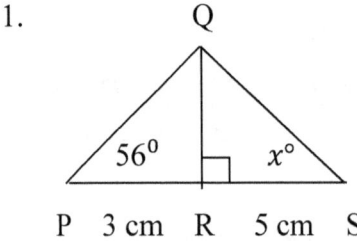

P   3 cm   R   5 cm   S

2.

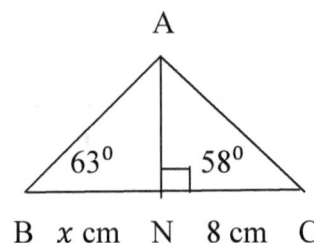

B x cm   N   8 cm   C

3.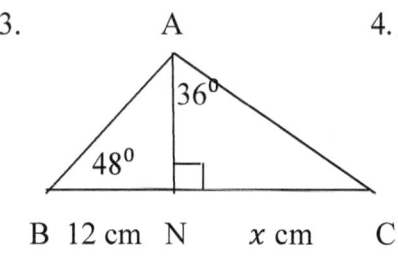

B 12 cm N   x cm   C

4.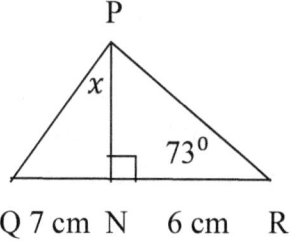

Q 7 cm N   6 cm   R

5.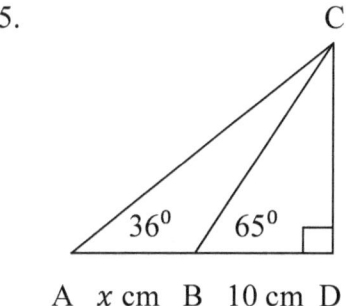

A   x cm B   10 cm D

6.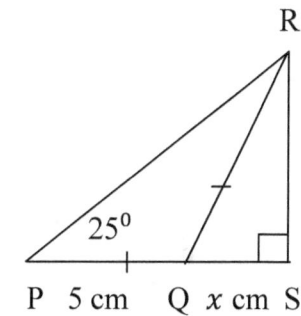

P   5 cm   Q x cm S

Angle of Elevation

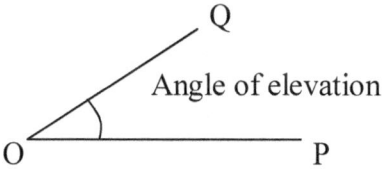

Figure 1.3

A boy whose eye is at O turns his eyes upwards to look at an object at Q on top of a building. In order to look up at the object, he must turn his eye through angle POQ. The angle between the line OQ and the horizontal line OP is called the angle of elevation of Q from O. Notice that the angle of elevation to an object is the angle formed by the line of sight to the object and a horizontal line.

Example

At a certain time of a day the angle of elevation of the sun is 63^0. Find the length of the shadow cast by a building 30 metres high. Give your answer to one decimal place.

Draw a diagram that represents the given information.

Let x represent the length of the shadow.

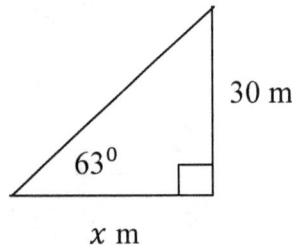

30 m

63^0

x m

From the diagram, we have

$$\tan 63° = \frac{30}{x}$$

$$x = \frac{30}{\tan 63°}$$

$$= 15.3$$

The length of the shadow is 15.3 m

Try this 7

The angle of elevation of the top of a tower from a point on the ground is 48^0. If the foot of the tower is 150 metres from the point, find the height of the tower. Give your answer correct to 1 decimal place

Angle of Depression

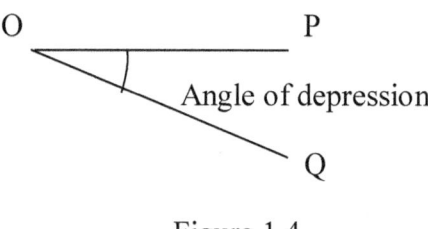

Figure 1.4

In Figure 1.4, Q is an object at a lower level than the eye level. In order to look at the object at Q, you will have to turn your eyes downwards through angle POQ. The angle between the line OQ and the horizontal line OP is called the angle of depression of Q from O. Notice that the angle of depression is the angle formed by the line of sight and a horizontal line.

Example

From the top of a tower the angle of depression of a building is 22^0. If the distance of the foot of the tower from the building is 185 metres, find the height of the tower. Give your answer to the nearest metre.

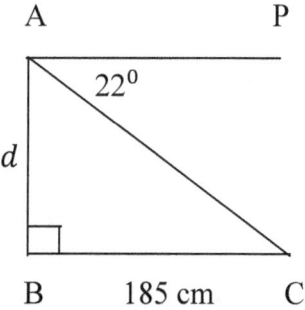

Let d represent the height of the tower. The line AP and BC are parallel, so $\angle CAP = \angle BCA$. Therefore we have $\angle BCA = 22°$.

Notice that the angle of depression from the top of the tower to the building equals the angle of elevation from the building to the top of the tower.

From $\triangle ABC$

$$\frac{d}{185} = \tan 22°$$

$$d = 185\tan 22°$$

$$= 185 \times 0.404$$

$$= 75$$

The height of the tower is 75 m

Try this 8

From the top of a tower the angle of depression of a point on the ground is 25^0. If the tower is 120 m high, find the distance from the point to the foot of the tower. Give your answer to the nearest metre.

Exercise 1.1(e)

1. From a point on the ground the angle of elevation of the top of a pole is 60^0. If the distance of the point from the foot of the pole is 30 m calculate the height of the pole.

2. From a point on the ground the angle of elevation of the top of a tower is 30^0. If the distance of the point from the foot of the tower is 300 m, calculate the height of the tower.

3. From a point on the ground the angle of elevation of the top of a tower is 27^0. If the height of the tower is 100 m, calculate the distance of the point from the foot of the tower.

4. From a point on the ground the angle of elevation of the top of a pole is 38^0. If the height of the pole is 120 m, calculate the distance of the point from the foot of the pole.

5. From the top of a tower the angle of depression of a point on the ground is 18^0. If the distance of the point from the foot of the pole is 150 m, calculate the height of the pole.

6. From the top of a cliff the angle of depression of a boat is 13^0. If the distance of the boat from the foot of the cliff is 250 m, calculate the height of the cliff.

7. From the top of a tower the angle of depression of a building is 25^0. If the distance of the building from the foot of the tower is 300 m, calculate the height of the tower.

8. From a 60 m tower on a coast, the angle of depression of a boat is 12^0. How far is the boat from the tower?

9. From a cliff 75 m high, the angle of depression to a ship is 16^0, how far is the ship from the cliff?

10. An aircraft flying at 1200 m above the ground is seen from the air port at an angle elevation of 12^0. How far away is the aircraft from the air port?

11. A man is 100 m from a building. He finds that the angle of elevation to the top of the building is 23^0. If the man's eye level is 1.55 m above the ground, calculate the height of the building.

12. A man is standing 120 m from a bridge. He determines that the angle of elevation to the top of the bridge is 35^0. The man's eye level is 1.45 m above the ground. Calculate the height of the bridge, correct to nearest metre.

13. From two points on the opposite sides of a tower 20 m high, two men observes its angles of elevation to be 25^0 and 34^0. How far are the men apart?

14. Two boys on opposite sides of a vertical pole are 20 metres apart. Each boy measures the angle of elevation of the top of the pole. If the angles are 32^0 and 45^0, calculate the height of the pole to the nearest tenth.

15. At a point P, the angle of depression of a point Q on an airfield from an airplane flying on a straight horizontal course at 120 km h $^{-1}$ was 15^0. If the plane was vertically over Q in 10 minutes, at what height was the plane flying?

16. A 12.5 m flag pole is placed on top of a tall building. A man standing directly in front of the building and flag pole, measures the angle of elevation to the bottom of the flag pole to be 43.5^0 and to the top of the flag pole to be 49.2^0. Calculate the height of the building to the nearest metre.

17. From a point A, a boy measured the angle of elevation of the top T of a tower to be 41^0. He then moved 30 m closer to the tower to a point B. From the point B he measured the angle of elevation to the top of the tower to be 52^0. Calculate the height of the tower to the nearest metre.

1.2 Cosine

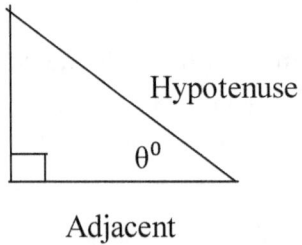

Adjacent

The cosine of the angle θ, written as $\cos \theta$ is defined as

$$\cos \vartheta° = \frac{length\ of\ adjacent\ side}{length\ of\ hypotenuse}$$

Try this 9

The triangles shown below are right-angled triangles. Read off the cosine of each acute angle in these triangles without simplifying

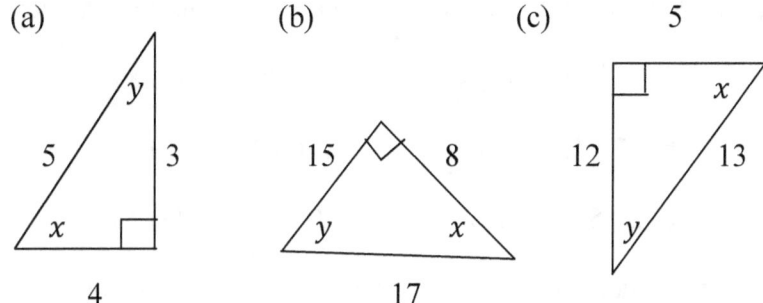

(a)   (b)   (c)   5

Using the Calculator

Examples

Find cos 65.8°

Enter [COS] [65.8⁰] =

You should be given the answer 0.4099

Try this 10

Find cos 79.3°

The cosine of an angle is 0.6947 find the size of the angle

Enter: [SHIFT] [COS] [0.6947] =

You should be given the answer 46.0^0 (1 d.p.)

Try this 11

The cosine of an angle is 0.4540 find the size of the angle.

Exercise 1.2(a)

Find correct to 4 significant figures the cosine of the following angles

1. (a) 7^0 (b) 21^0 (c) 49^0 (d) 76^0 (e) 87^0

2. (a) 3.6^0 (b) 37.5^0 (c) 52.3^0 (d) 68.2^0 (e) 79.4^0

3. (a) 16.25^0 (b) 28.53^0 (c) 41.76^0 (d) 62.18^0 (e) 85.36^0

Find correct to 1 decimal place the size of an angle if cosine of the angle is:

4. (a) 0.9962   (b) 0.8090   (c) 0.5592   (d) 0.3090   (e) 0.1219

5. (a) 0.9593   (b) 0.8957   (c) 0.7120   (d) 0.5120   (e) 0.1788

6. (a) 0.9535   (b) 0.7811   (c) 0.4916   (d) 0.2001   (e) 0.0099

Using the Cosine Ratio

Examples

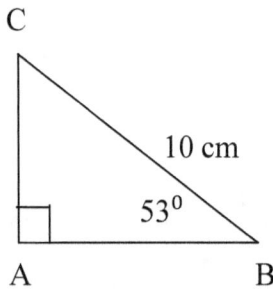

In the diagram triangle ABC is a right-angled triangle. $\angle BAC = 90°$, $\angle ABC = 53°$ and BC = 10 cm. Calculate the length of AB to the nearest unit

You can see from the diagram that AB is the adjacent side and BC is the hypotenuse.

$$\frac{AB}{10} = \cos 53°$$

$$AB = 10 \cos 53°$$

$$= 6 \text{ cm}$$

Try this 12

Sketch a right-angled triangle with $\angle PQR = 90°$, $\angle QRP = 36°$, QR = 14 m. Find the length of PR, correct to one decimal place.

Example

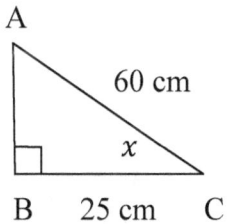

A

60 cm

x

B   25 cm   C

In the diagram ABC is a right-angled triangle. BC = 25 cm, AC = 60 cm $\angle ABC = 90°$ and $\angle BCA = x°$. Find size of angle x, correct to the nearest degree

$$\cos x° = \frac{25}{60}$$

$$= 0.4167$$

$$x = 65°$$

Try this 13

Sketch a right-angled triangle with $\angle PQR = 90°$, $\angle QRP = x°$, $QR = 8$ m and $PR = 15$ m. Find the size of angle x, correct to one decimal place.

Exercise 1.2(b)

The triangles below are right-angled triangles. Calculate the length of the side or the size of the angle marked x in the following diagrams.

1(a)   (b)   (c)

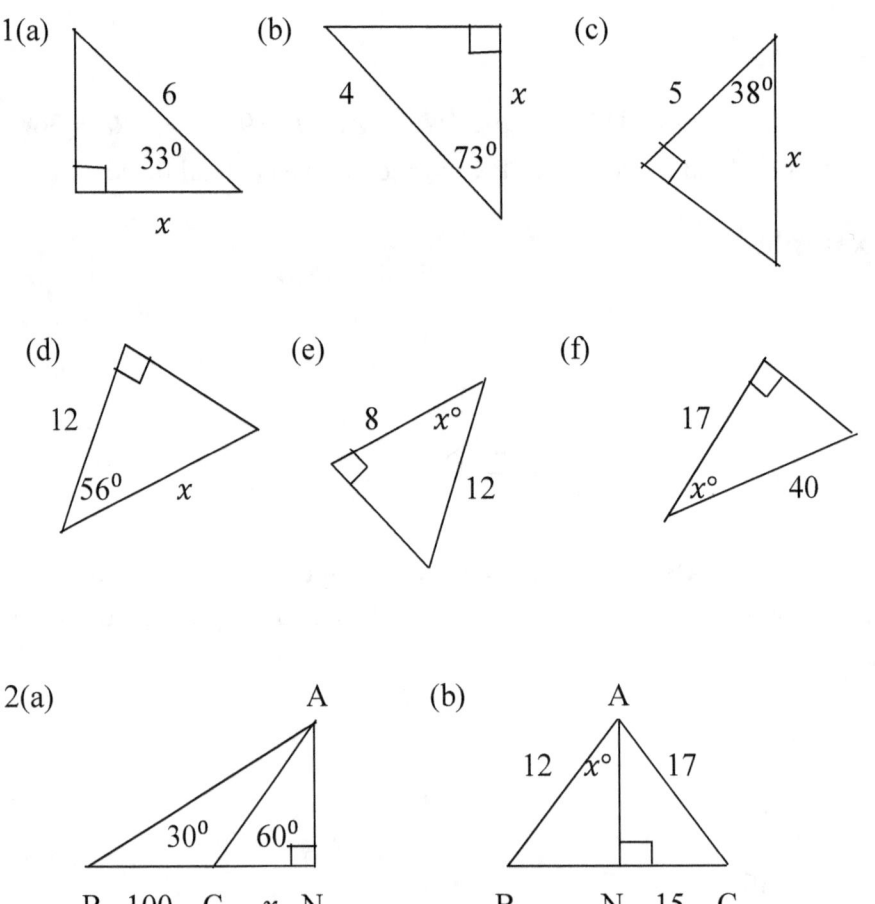

(d)   (e)   (f)

2(a)   (b)

(c)

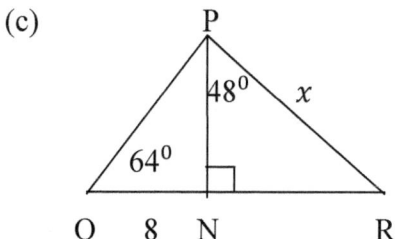

1.3 Sine

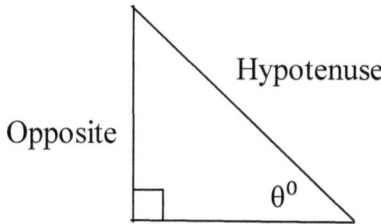

The sine of an angle, written as sin θ is defined as

$$\sin \theta° = \frac{length\ of\ opposite\ side}{length\ of\ hypotenuse}$$

Try this 14

The triangles shown below are right-angled triangles. Read off the sine of each acute angle in these triangles without simplifying

(a)   (b)   (c)

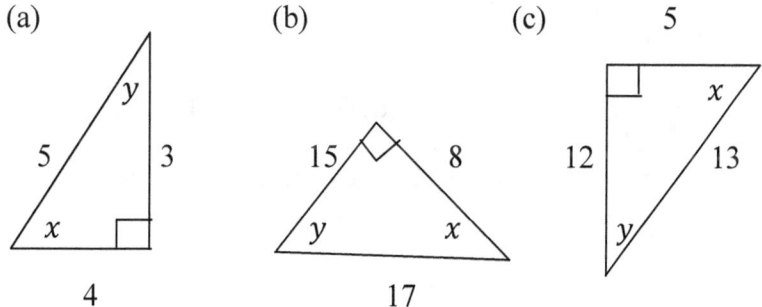

Using the Calculator

Example

Find sin 38.7°

Enter: [SIN] [38.7⁰] =

You should be given the answer 0.6252

Try this 15

Find sin 54.9°

The sine of an angle is 0.6691 find the size of the angle

Enter: [SHIFT] [SIN] [0.6691] =

You should be given the answer 42.0^0 (1 d.p.)

Try this 16

The sine of an angle is 0.9563 find the size of the angle.

Exercise 1.3(a)

Find correct to 4 decimal places the sine of each of the following angles

1. (a) 2^0 (b) 23^0 (c) 41^0 (d) 57^0 (e) 81^0

2. (a) 1.6^0 (b) 12.7^0 (c) 37.8^0 (d) 63.5^0 (e) 82.3^0

3. (a) 24.76^0 (b) 43.27^0 (c) 59.14^0 (d) 71.38^0 (e) 87.05^0

Find correct to 1 decimal place an angle if the sine of the angle is:

4. (a) 0.0872   (b) 0.6157   (c) 0.7431   (d) 0.9397   (e) 0.9925

5. (a) 0.0070   (b) 0.4446   (c) 0.6769   (d) 0.9650   (e) 0.9988

6. (a) 0.1057   (b) 0.5320   (c) 0.8227   (d) 0.9234   (e) 0.9800

Using the Sine Ratio

Examples

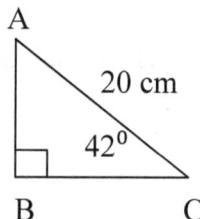

In the diagram above ABC is a right-angled triangle. AC = 20 cm, $\angle ABC = 90°$ and $\angle BCA = 42°$. Find the length of AB, correct to the nearest centimetre.

You can see from the diagram that AB is the opposite side to angle BCA and AC is the hypotenuse.

$$\frac{AB}{20} = \sin 42°$$

$$AB = 20 \sin 42°$$

$$= 20 \times 0.6691$$

$$= 13 \text{ cm}$$

Try this 17

Sketch a right-angled triangle with $\angle ABC = 90°$, $\angle BCA = 56°$ and AB = 8 m, find AC, correct to 1 decimal place.

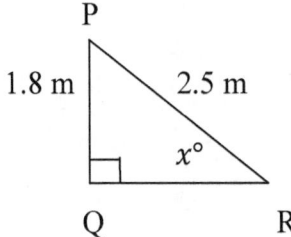

In the diagram above PQR is a right-angled triangle. $\angle PQR = 90°$ and $\angle QRP = x°$, PQ = 1.8 m and PR = 2.5 m. Find the size of angle x, correct to the nearest degree.

$$\sin x° = \frac{1.8}{2.5}$$

$$= 0.72$$

$$x = 46°$$

Try this 18

Sketch a right-angled triangle with $\angle ABC = 90°$, $\angle CAB = x°$, BC = 20 cm and AC = 25 cm, find the size of angle x, correct to 1 decimal place.

Exercise 1.3(b)

The triangles below are right-angled triangles. Calculate the length of the side or the size of the angle marked x in each of the following diagrams.

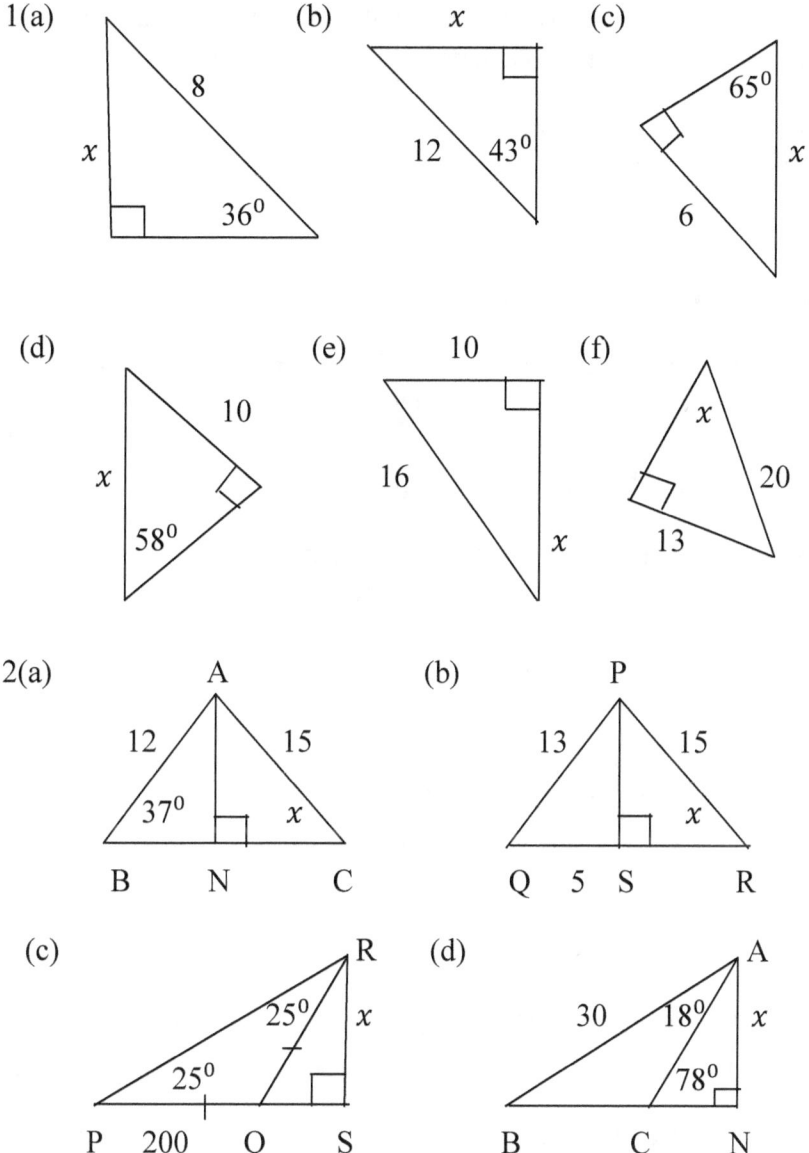

1(a)   (b)   (c)

(d)   (e)   (f)

2(a)   (b)   (c)   (d)

Questions involving trigonometry and right angled triangles will only require you to use the sine, cosine and tangent ratios. Which ratio you use will depend on the information given in the question. There are couple of ways to help you remember which ratio to use. It may help to use the mnemonic SOHCAHTOA derived from the ratios

$$sin = \frac{opp}{hyp}, cos = \frac{adj}{hyp} \text{ and } tan = \frac{opp}{adj}$$

Special Triangles

There are two special triangles that arise in trigonometry: the 30^0 - 60^0 - 90^0 triangle and the 45^0 - 45^0 - 90^0 triangle.

Trigonometric Ratios for 45^0

The triangle shown below is a right-angled isosceles triangle. The equal sides are of length 1 as shown in Figure 1.5

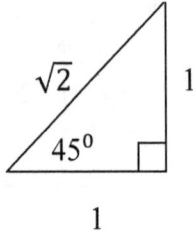

Figure 1.5

The Pythagoras' theorem gives the length of the hypotenuse:
$$\sqrt{1^2 + 1^2} = \sqrt{2}$$

You can see from Figure 1.5 that

$$\sin 45° = \frac{1}{\sqrt{2}} = \frac{\sqrt{2}}{2}$$

$$\cos 45° = \frac{1}{\sqrt{2}} = \frac{\sqrt{2}}{2}$$

and $\tan 45° = 1$

Trigonometric Ratios for 30⁰ and 60⁰

The triangle shown below is an equilateral triangle of length 2. An altitude drawn from any vertex bisects the angle at the vertex and it also bisect the base as shown in Figure 1.6

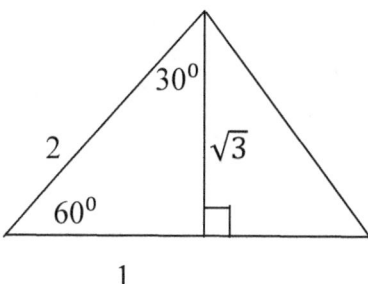

Figure 1.6

The Pythagoras' theorem gives the length of the altitude:
$$\sqrt{2^2 - 1} = \sqrt{3}$$

You can see from Figure 1.6 that

$$\sin 60° = \frac{\sqrt{3}}{2} \qquad \text{and} \qquad \sin 30° = \frac{1}{2}$$

$$\cos 60° = \frac{1}{2} \qquad\qquad \cos 30° = \frac{\sqrt{3}}{2}$$

$$\tan 60° = \sqrt{3} \qquad\qquad \tan 30° = \frac{1}{\sqrt{3}} = \frac{\sqrt{3}}{3}$$

Example

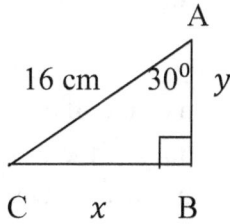

In $\triangle ABC$, $\angle ABC = 90°$, $\angle BAC = 30°$ and $AC = 16$ cm. Find without using tables or calculator the values of x and y

$$\frac{x}{16} = \sin 30°$$

$$x = 16 \times sin30°$$

$$= 8 \text{ cm}$$

And $\frac{y}{16} = cos30°$

$$y = 16 \times cos30°$$

$$= 8\sqrt{3} \text{ cm}$$

Try this 19

A ladder leans against a vertical wall with which it makes an angle of 30^0. If the ladder reaches the bottom of a window that is 15 metres above the ground, find without using tables or calculator the length of the ladder. Leave surd in your answer.

Exercise 1.3(c)

You should answer Exercise 1 – 4, without using tables or calculator

1. A ladder 17 metre long rest against a wall and makes an angle of 30^0 with the wall. How far is the foot of the ladder from the wall?

2. A boy is flying a kite on a 50 metre string. The string makes an angle of 60^0 with the ground. Find the height of the kite above the ground.

3. Two men stand on the same side of a 36 metre tower. From their location the angles of elevation of the top of the tower are 30^0 and 60^0 respectively. What is the distance between the two men?

4. Two trees are on a level ground. At a point on the ground midway between then, the angles of elevation of their tops are 60^0 and 30^0 respectively. The height of the taller tree is 15 metres. Find the height of the other tree.

Sometimes you need not find the value of a trigonometric ratio from a table or calculator.

Example

Given $\cos A = \frac{5}{13}$ and $0° < A < 90°$, find $\sin A$ and $\tan A$

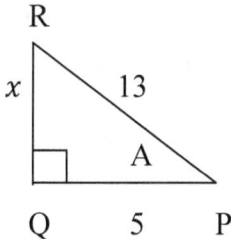

Since cosine is $\frac{length\ of\ adjacent\ side}{length\ of\ hypotenuse}$, $PQ = 5$ and $PR = 13$. Using the Pythagoras' theorem we have

$x^2 = 13^2 - 5^2$

$\quad = 169 - 25$

$\quad = 144$

$x = 12$

From the diagram, you can see that

$\sin A = \frac{12}{13}$ and $\tan A = \frac{12}{5}$

Try this 20

Given $\tan A = \frac{15}{8}$ and A is an acute angle find $\sin A$ and $\cos A$.

Exercise 1.3(d)

The triangles below are right-angled triangles. Calculate the length of the side or the size of the angle marked x in each diagram.

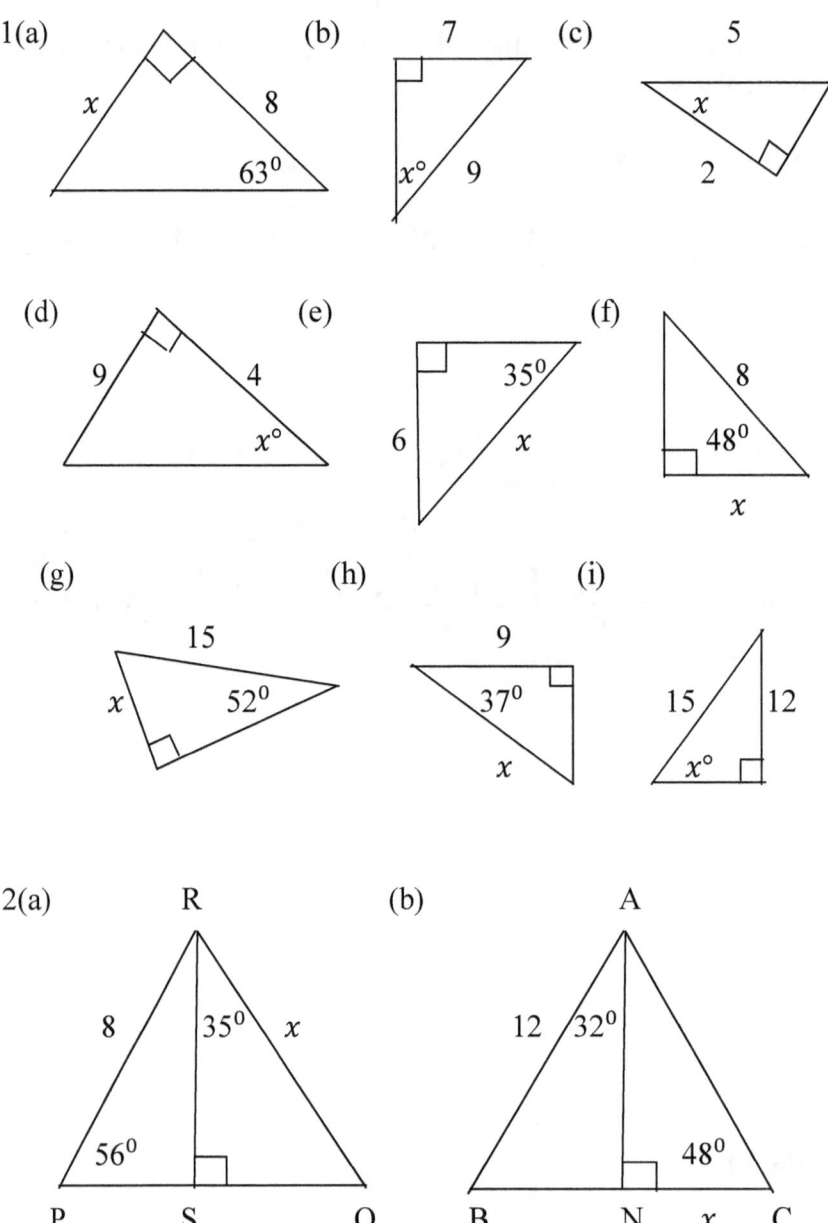

1(a)

(b)   7   (c)   5

(d)   (e)   (f)

(g)   (h)   (i)

2(a)   R   (b)   A

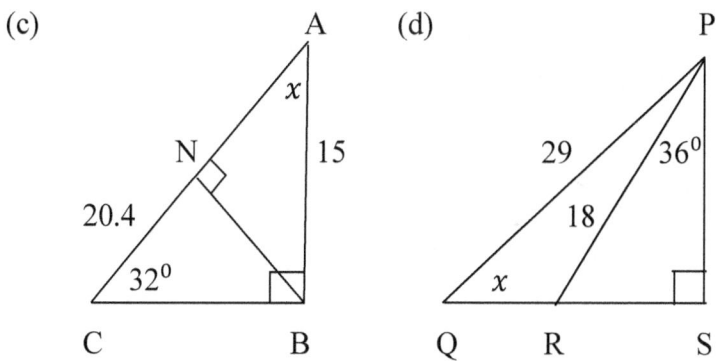

(c)

(d)

3. If 25 sin A = 7 and A is an acute angle find tan A and cos A

4. If $\cos A = \frac{12}{13}$ and $0° < A < 90°$, find the value of $\frac{2\tan A}{1-\sin A}$

5. Given $\tan A = \frac{8}{15}$ and $0° < A < 90°$, find the value of $1 - 2sin^2A$

6. Given $\cos A = \frac{5}{13}$ and $0° < A < 90°$, find $1 + tan^2A$

7.

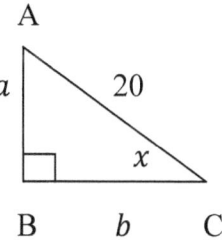

In the diagram $\angle ABC = 90°$, AC = 20 and $\tan x° = \frac{3}{4}$, find the values of a and b without using tables or calculator.

8. In the diagram ABCD is a rectangle AF = 12 cm, AB = 16 cm, FD = 15 cm and $\angle BFA = x°$. Find the length of EC without using tables or calculator

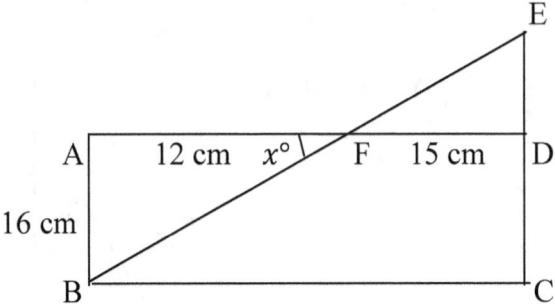

1.4 Solving Problems using right-angled triangles

In solving a triangle problem, you will find it useful to draw a diagram that represents the given information. On your diagram indicate the given measurements and the measurements you are to find.

Example

A ladder 2 m long rest against a wall so that it reaches a window above the ground. The ladder is inclined at 25^0 to the wall. Find the height of the window above the ground.

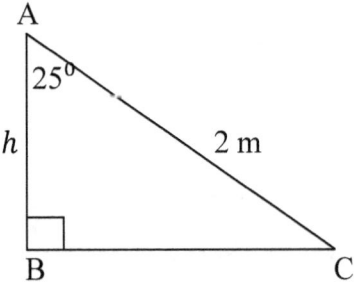

In the diagram, AB represents the wall, BC represents the ground and AC represents the ladder. Let h be the height of the window above the ground.

$$\frac{h}{2} = \cos 25°$$

$$h = 2\cos 25°$$

$$= 2 \times 0.9063$$

$$= 1.8 \text{ m}$$

The window is 1.8 m above the ground

Try this 21

To secure a pole 200 m long against high wind a rope is attached to a ring 5 metre from the top of the pole. If the rope forms a $15°$ angle with the pole, calculate the distance from the pole to the rope anchor in the ground.

Exercise 1.4

1. A ladder 2.5 m long just reaches the top of a wall. If the foot of the ladder is 1.2 m from the wall, find the size of the angle between the ladder and the wall.

2. A ladder 3.6 m long is placed against a wall at an inclination of $46°$ to the ground. How high above the ground is the top?

3. A ladder leans against a vertical wall with which it makes an angle of $38°$. The foot of the ladder is 2 m from the wall. Find its length.

4. A pole is held in a vertical position by a rope 1.8 m long fastened from its top to a peg in the ground. Calculate the height of the pole if the angle between the rope and the pole is $23°$.

5. A boy is flying a kite on a 45- metre string. The string is making a $50°$ angle with the ground. How high above the ground is the kite?

6. A boy stands at A on a river bank directly opposite a pole B on the other bank. His distance from a pole at C on the same bank as B is 100 m. Find the width of the river if $\angle BCA$ is 52^0.

7. An aircraft takes off from a run way and rises at an angle of 15^0 to the horizontal. Find the height of the aircraft at a distance 1200 m from the place where it left the ground.

8. An aircraft flying at 480 km h^{-1} climbs at an angle of 6^0 to the horizontal. What will be its height after flying for 30 minutes?

9. A car travels 200 m up a hill which slopes at 13^0 to the horizontal. How high is the car above its starting point?

10. A plank 360 cm long rests with one end on horizontal ground and the other end on the top of a vertical wall 180 cm high. Calculate the angle which the plank makes with the ground.

1.5 Problems in Three -Dimension

In the previous section we considered trigonometry problems in one dimension. We can also apply the same trigonometric ratios to 3-dimensional problems, provided we can identify some right-angled triangles to work with. The general method is illustrated in the example below.

Example

A boy who was 3 m east of a building walked 4 m south. From his final position the angle of elevation of the building is 58^0. Find the height of the building.

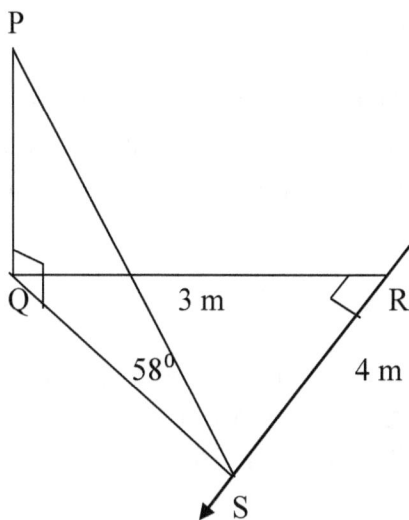

Let R be the first position of the boy and S his final position. PQ is the height of the building. In triangle QRS,

$$QS^2 = QR^2 + RS^2$$

$$= 3^2 + 4^2$$

$$= 9 + 16$$

$$= 25$$

$$QS = 5 \text{ m}$$

In triangle PQS

$$\frac{PQ}{QS} = \tan 58°$$

$$PQ = 5 \tan 58°$$

$$= 5 \times 1.6$$

$$= 8 \text{ m}$$

The height of the building is 8 m.

Try this 22

A horizontal lawn is in the shape of a right angled triangle ABC with AB = 30 m, BC = 40 m and $\angle ABC = 90°$. At A stands a vertical pole AP. From B the angle of elevation of P is 50.2^0. Find the angle of elevation of C from P.

Exercise 1.5

1. A horizontal play ground ABCD is in the shape of a rectangle with AB = 8 m, BC = 6 m. At C stands a vertical pole CP which is 3.5 m high. Find the angle of elevation of P from A.

2. A right angled triangle PQR is marked on a horizontal ground. PQ = 3 m, QR = 4 m and $\angle PQR = 90°$. A vertical post AP, 12 m high stands at P and a taut wire joins A to R. Find

(i) the size of angle PRA   (ii) the length of AR

3. A ship is 90 m due west of a cliff 170 m high. What will be the angle of elevation of the top of the cliff from the ship when the ship has sailed 120 m due south?

4. From a point A, the angle of elevation of an aircraft 1600 m high and due north of A is 45^0. B is a point 2000 m due east of A and at the same level as A. Find the angle of elevation of the aircraft from B.

5. A is the top of a tower AB, 120 m high, which stands on a horizontal plane. From P, a point in the plane due West of B, the angle of elevation of A is 30^0. From Q, another point in the plane which is due south of B, the angle of elevation of A is 60^0. Find the distance PQ.

6. A shore light 8 m above sea-level, can be seen from two ships A and B, the angles of elevation from A and B being 58^0 and 45^0

respectively. The light is due west of A and B is south of A. Find the distance between the ships. Give your answer to the nearest unit.

7. A level straight road is 1.6 m wide. At two points A and B, 1.2 m apart on the same side of the road, two equal vertical posts AP and BQ are erected. From a point C at ground level on the other side of the road and directly opposite A, the angle of elevation of P is 37^0. Calculate the angle of elevation of Q from C.

8. A vertical pole AB stands at a point B on a horizontal plane. From P, due East of the pole, the tangent of the angle of elevation of A is $\frac{1}{5}$; from Q due South of the pole, the tangent of the angle of elevation of A is $\frac{1}{12}$. If PQ is 65 metres, find the height of the pole, without using tables or calculator.

1.6 Lengths in Circles

A straight line joining the centre of a circle to the midpoint of a chord

A straight line which joins the centre of a circle to the middle point of a chord is perpendicular to the chord.

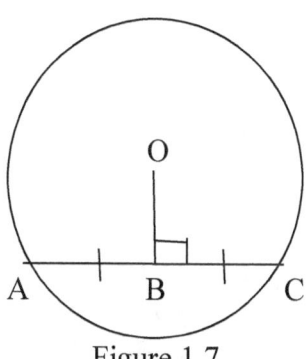

Figure 1.7

In Figure 1.7, O is the centre of the circle and AB = BC. Therefore $\angle OBA = \angle OBC = 90°$.

A straight line through the centre of a circle perpendicular to a chord

A perpendicular drawn from the centre of a circle to a chord bisect the chord

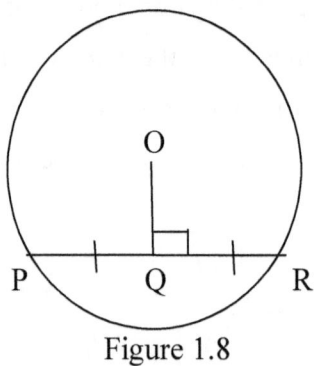

Figure 1.8

In Figure 1.8, O is the centre of the circle and $\angle OQR = 90°$. Thus, PQ = QR

Examples

The chord of a circle is 16 cm long. It is 6 cm from the centre of the circle. Find the radius of the circle.

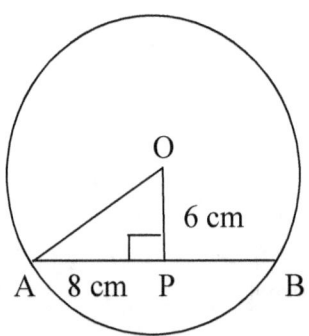

In the diagram, O is the centre of the circle, AB = 16 cm and OP = 6 cm. $\angle OPA = 90°$. Because AP = PB, AP = 8 cm. By the Pythagoras' theorem

$$OA^2 = AP^2 + OP^2$$

$$= 8^2 + 6^2$$

$$= 64 + 36$$

$$= 100$$

$$OA = 10 \text{ cm}$$

The radius of the circle is 10 cm.

Try this 23

A circle has a radius 5 cm. A chord is drawn in the circle at a distance of 4 cm from the centre. Find the length of the chord.

A circle has centre O and radius 10 cm. A chord AB is drawn which is 6 cm from the centre. Find size of angle AOB.

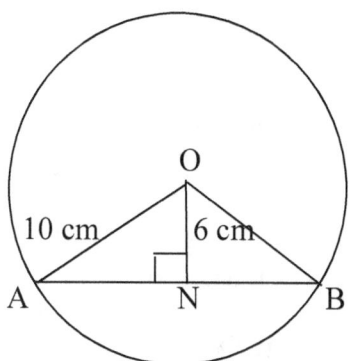

$$\cos \angle AON = \frac{6}{10}$$

$$= 0.6$$

$$\angle AON = 53°$$

Because triangle AOB is an isosceles triangle, $\angle AON = \angle BON$

$$\angle AOB = 2\angle AON$$

$$= 2 \times 53°$$

$$= 106°$$

Try this 24

A circle has centre O and radius 6 cm. A chord AB is drawn which is 4 cm from the centre. Find the size of angle AOB.

Exercise 1.6(a)

1. AB is a chord of a circle of radius 10 cm and AB = 16 cm, find the distance of the centre of the circle from AB.

2. A chord of length 10 cm is at a distance of 12 cm from the centre of the circle, find the radius

3. A chord of a circle of radius 15 cm, is at a distance of 9 cm from the centre, find its length.

4. A chord AB of a circle is 10 cm long and it subtends an angle of 64^0 at the centre O. Find the distance of O from AB.

5. A chord AB of a circle radius 5 cm subtends an angle of 86^0 at the centre O. Calculate the length of the chord.

6. AB is a chord of a circle, centre O and radius 8 cm. AB = 10 cm, find $\angle AOB$.

The Tangent to a Circle

Recall that the tangent to a circle is a straight line that touches the circle at exactly one point, and the radius drawn to the point of contact of the tangent is perpendicular to the tangent.

Tangent from external point

1. Two tangents drawn to circle from an external point are equal and subtend equal angles at the centre.

2. The line joining the centre to the external point bisects the angle between the tangents

Example

AT and BT are two tangents to a circle centre O at A and B respectively. If the radius of the circle is 3 cm and $\angle ATB = 60°$, find (a) OT and (b) $\angle AOB$

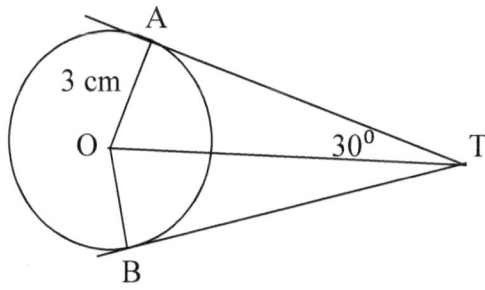

OT bisects $\angle ATB$. Thus, $\angle OTA = 30°$

$$\sin 30° = \frac{3}{OT}$$

$$OT = \frac{3}{\sin 30°}$$

$$= 6 \text{ cm}$$

(b) $\angle AOT = 90° - 30°$

$\quad\quad = 60°$

Thus, $\angle AOB = 120°$

Try this 25

An arc AB of a circle centre O subtends an angle of 76^0 at the centre. AT and BT are the tangent to the circle at A and B. Find the radius of the circle if the length of the tangent is 5 cm.

Exercise 1.6(b)

1. AT is tangent to a circle, centre O at A. If the radius of the circle is 4 cm and $\angle AOT = 63°$, find the length of AT.

2. AT and BT are tangents to a circle, centre O and radius 6 cm. If OT = 8 cm find $\angle AOB$.

3. AT and BT are tangents to a circle, centre O. If the radius is 8 cm and OT = 12 cm find $\angle ATB$.

4. AT and BT are tangents to a circle, centre O and radius 5 cm. If OT = 13 cm find the length of AT.

5.

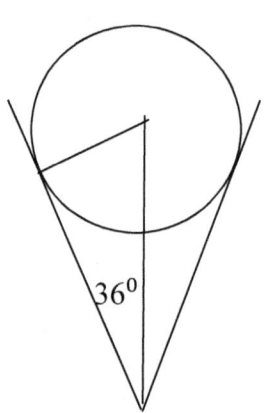

36^0

The diagram above shows a spherical ball, resting inside a hallow cone of semi-vertical angle 36^0, which stands with its vertex downwards. Find the radius of the ball if the height of the centre of the ball above the vertex of the cone is 12 cm.

6.

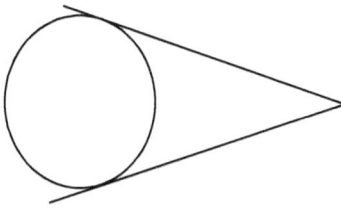

The diagram above shows a piece of string passing tightly round a circular plate and a pin placed 10 cm from the centre of the plate. Find the length of the string if the radius of the plate is 6 cm.

1.7 Sine and Cosine Rule

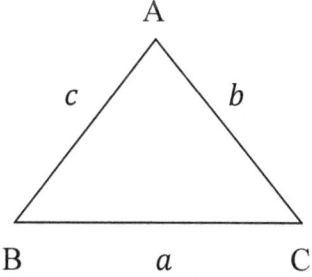

The sides of the triangle ABC are labelled as shown in the diagram above. Notice that a, b and c represents the sides opposite angles A, B and C respectively.

Sine Rule

If ABC is a triangle with a, b and c representing the length of sides opposite angles A, B and C respectively, then

$$\frac{a}{\sin A} = \frac{b}{\sin B} = \frac{c}{\sin C}$$

The rule can be written as:

$$\frac{\sin A}{a} = \frac{\sin B}{b} = \frac{\sin C}{c}$$

The sine rule can be used to solve a triangle when given:

1. the sizes of two angles and the length of any side

2. the lengths of two sides and the size of the angle opposite one of these sides of the triangle

Example

Find the length of the side BC of triangle ABC given $B = 50°$, $C = 60°$ and $b = 8$ cm

Draw and label the triangle

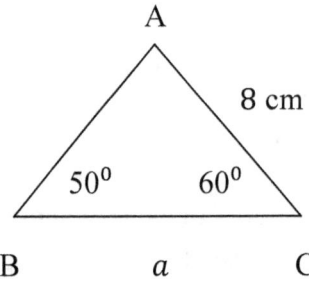

First find angle A.

$A + 50 + 60 = 180$

$$A = 70°$$

Next, use the sine rule

$$\frac{a}{\sin 70°} = \frac{8}{\sin 50°}$$

$$a = \frac{8\ sin70°}{\sin 50°}$$

$$= 9.8 \text{ cm}$$

The length of BC is 9.8 cm

Try this 26

Given triangle ABC with $B = 45°$, $C = 70°$, $a = 8$ cm, calculate the length of b.

Cosine Rule

If a, b and c represents the sides opposite angles A, B and C respectively, then

$$a^2 = b^2 + c^2 - 2bc\cos A$$

Two other forms are:

$$b^2 = a^2 + c^2 - 2ac\cos B$$

$$c^2 = a^2 + b^2 - 2ab\cos C$$

The cosine rule can be used to solve a triangle when given:

1. the lengths of two sides and the size of the included triangle

2. the lengths of the three sides

Example

Given triangle ABC with $A = 60°$, $B = 8$ cm and $c = 6$ cm, find the length of a

First, draw and label the triangle

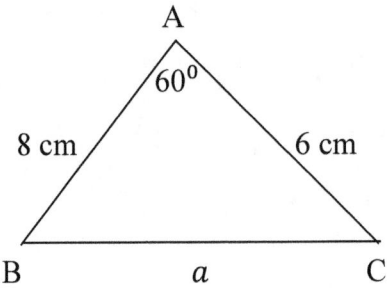

$$a^2 = b^2 + c^2 - 2bc\cos A$$

$$= 6^2 + 8^2 - 2(6)(8)\cos 60°$$

$$= 52$$

$$a = 7.2 \text{ cm}$$

Try this 27

Given triangle ABC with $a = 5$ cm, $b = 7$ cm and $C = 76°$ find the length of c

Solution of Triangles

A triangle has three sides and three angles. A triangle is said to be solved when all three sides and all the three angles are known.

The trigonometric ratios cannot be used to solve triangles which are not right-angled. When at least two sides and an angle are given you can use the sine rule and the cosine rule to find three unknown quantities.

Solving Triangles

1. Given one side and two angles of triangle ABC

   (a) Find the third angle from $A + B + C = 180°$

   (b) Use the sine rule to find the other two sides

2. Given two sides and an angle of triangle ABC

   (a) Use the sine rule to find one of the two remaining angles

   (b) Find the third angle from $A + B + C = 180°$

   (c) Use sine rule to find the third side

3. Given two sides and the included angle of triangle ABC

   (a) Use the cosine rule to find the third side

   (b) Use the sine rule to find the smaller of the two remaining angles

   (c) Find the third angle from $A + B + C = 180°$

4. Given three sides of triangle ABC

   (a) Use the cosine rule to find the smallest angle

   (b) Use the sine rule to find the smaller of the two remaining angles

   (c) Find the third angle from $A + B + C = 180°$

Example

Solve triangle ABC, given $a = 6$, $b = 9$, $C = 85°$. Give your answers to the nearest units

First, draw and label the triangle

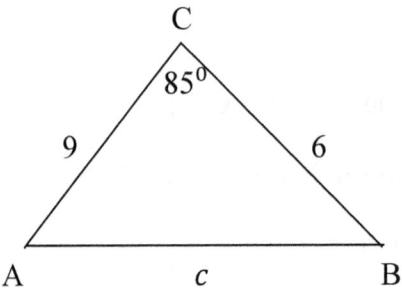

Since two sides and the included angle of the triangle is given, use the cosine rule

$$c^2 = a^2 + b^2 - 2ab\cos C$$

$$= 6^2 + 9^2 - 2(6)(9)\cos 85°$$

$$c = 10.4$$

Next, use sine rule to find A

$$\frac{sine\ A}{6} = \frac{\sin 85°}{10.4}$$

$$\sin A = 0.5747$$

$$A = 35°$$

Find B from $A + B + C = 180$

$$35 + B + 85 = 180$$

$$A = 60°$$

Try this 28

Solve triangle ABC, given $b = 5$, $c = 4$, $A = 60°$. Give your answer to the nearest tenth

Exercise 1.7

1. Solve each triangle. Give your answers to the nearest tenth.

(a) $a = 4.8$, $B = 110°$, $A = 22°$ (b) $b = 15$, $C = 130°$, $A = 20°$

(c) $c = 12$, $A = 75°$, $B = 45°$ (d) $a = 3$, $b = 5$, $C = 20.5°$

(e) $b = 5$, $c = 3$, $A = 48.2°$ (f) $C = 37°$, $a = 13.5$, $A = 68°$

(g) $a = 12$, $b = 18$, $c = 21$ (h) $a = 16$, $b = 14$, $c = 10$

2. The longest side of a triangle is 32 metres. The sizes of two angles of the triangle are 40^0 and 65^0. Find the length of the other two sides.

3. A house is built on a triangular plot of land. Two sides of the plot are 150 metres long and they meet at an angle of 85^0. If a fence is to be built around the property, how much fencing material is needed?

4. A ship is sighted at sea from two observation points on the coastline that are 30 kilometres apart. The size of the angle formed by the coastline and the line between the ship and the first observation point is 35^0. The size of the angle formed by the coastline and the line between the ship and the second observation points is 46^0. How far is the ship from the first observation point?

5. Two sides of a triangular plot of land have lengths of 40 metres and 60 metres. The size of the angle formed by those sides is 46.2^0. Find the perimeter of the plot to the nearest tenth.

6. The sides of a triangle are 50 metres, 65 metres and 72 metres long. Find the size of the angle opposite the longest side to the nearest degree.

7. The hour hand of a clock is 9 cm long and the minute hand is 12 cm. How far are the tips of the hands at 4 o'clock?

8. A 50-metre television antenna stands on top of a building. From a point on the ground, the angles of elevation to the top and bottom of the antenna are 52^0 and 46^0 respectively. How tall is the building?

9. A ship at sea is 80 kilometres from one radio transmitter and 120 kilometres from another. The size of the angle formed by the signals is 110^0. How far apart are the transmitters?

10. A plane flew 800 kilometres north before turning 25^0 clockwise. It flew 1200 kilometres in that direction and then landed. How far is the plane from its starting point?

11. Two planes leave a port at the same time. One flies in the direction 60^0 east of north at a speed of 270 kilometres per hour. The other flies in the direction 40^0 east of south at a speed of 310 kilometres per hour. How far apart are the planes after 1 hour 30 minutes?

12. A pilot is flying from Accra airport to Kumasi, a distance of 270 kilometres. In order to avoid bad weather, he starts his flight 15 degrees off course and flies on this course for 15 minutes. If he flies at a speed of 300 kilometres per hour, how far is he from Kumasi?

Review exercise 1

1. Find the tangent of each of the following angles

(a) 37^0 (b) 58^0 (c) 65^0 (d) 72.6^0

2. Find the cosine of each of the following angles

(a) 8.5⁰   (b) 47⁰   (c) 62⁰   (d) 78.2⁰

3. Find the sine of each of the following angles

(a) 25⁰   (b) 63.8⁰   (c) 73⁰   (d) 85.4⁰

4. Find the size of an angle if the tangent of the angle is:

(a) 0.3564   (b) 0.7264   (c) 1.273   (d) 2.965

5. Find the size of the angle if the cosine of the angle is:

(a) 0.8763   (b) 0.7125   (c) 0.5260   (d) 0.2789

6. Find the size of the angle if the sine of the angle is:

(a) 0.4578   (b) 0.7809   (c) 0.8765   (d) 0.6549

7. Find the length of the side or the size of the angle marked x in each of the following diagrams. Give your answer correct to 1 d.p.

(a)

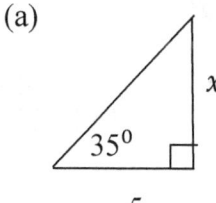

(b)

(c)

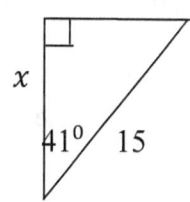

(d)

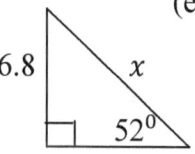

(e)

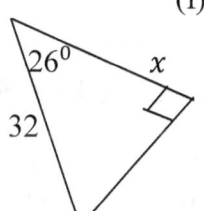

(f)

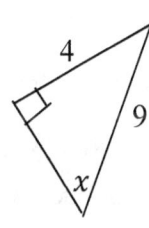

(g) 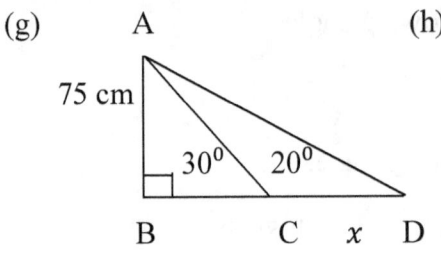   (h)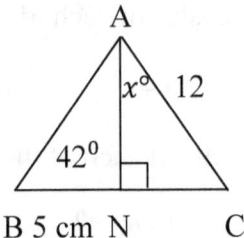

(i)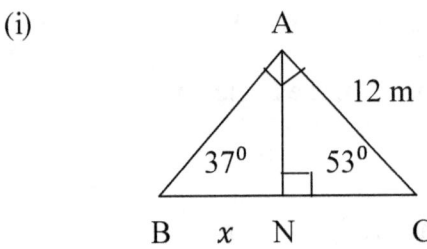

8. Given $\tan A = \frac{3}{4}$, find $\sin A$ and $\cos A$.

9. A ladder rest against the top of the perpendicular wall of a building and makes an angle of 63^0 with the ground. If the foot of the ladder is 1.5 m from the wall, calculate the height of the building

10. A pole stands on horizontal ground. At a point 80 m from the base of the pole, the angle of elevation of the top of the pole is 23^0. Calculate the height of the pole to the nearest metre.

11. From a point a man observes the angle of elevation of the top of a tower to be 19^0. He then moves 120 m nearer the tower and finds the angle of elevation is now 47^0. Determine the height of the tower.

12. From the top of a vertical cliff 80 m high the angle of depression of a boat is 20^0. Find the distance of the boat from the cliff.

13. From the top of a vertical cliff 90 m high the angles of depression of two ships lying due west of the cliff are 24^0 and 16^0 respectively. How far are the ships apart?

14. The elevation of a tower from two points, one due east of the tower and the other due west of it are 35^0 and 47^0 respectively. If the two points of observation are 250 m apart. Find the height of the tower to the nearest metre.

15. A communication mast is on top of a building. From a point 100 metres from the building the angles of elevation of the bottom and top of the mast are 15^0 and 23^0 respectively. How tall is the mast?

16. A ladder 3.5 m long just reaches the top of a wall. If the foot of the ladder is 2.0 m from the wall, find the angle between the ladder and the wall.

17. A ladder 2.5 m long is placed against a wall at an inclination of 36^0 to the ground. How high above the ground is the top?

18. AB is a chord of a circle of radius 12 cm. AB = 18 cm, find the distance of the centre of the circle from AB.

19. A chord AB of a circle radius 6 cm subtends an angle of 64^0 at the centre O. Find the length of the chord.

20. AT and BT are two tangents to a circle, centre O at A and B respectively. If the radius of the circle is 5 cm, and $\angle ATB = 80°$, find (a) OT and (b) $\angle AOB$

21. An arc AB of a circle centre O subtends an angle of 86^0 at the centre. AT and BT are the tangent to the circle at A and B. Find the radius of the circle if the length of the tangent is 8 cm

22. In triangle ABC $AB = 16$ cm, $BC = 25$ cm and $\angle ABC = 53°$. Calculate the length of AC, to the nearest tenth.

23. In triangle ABC $AB = 32$ m, $BC = 43$ m and $AC = 64$ m, calculate the size of angle CAB, to the nearest tenth.

24. Solve each triangle. Give your answers to the nearest tenth.

(a) $a = 3.5, B = 47°, C = 58°$ (b) $b = 10.7, c = 9.3, A = 61°$

(c) $A = 30°, a = 15, c = 20$ (d) $a = 8.9, b = 9.7, c = 17.0$

25. A ship is sighted at sea from two observation points on the coastline that are 1,250 metres apart. The size of the angles formed by the coastline and the lines between the ship and the observation points are 62^0 and 51^0 respectively. Calculate the distance of the observation points from the ship.

26. A tower stands on a horizontal plane. From a point in the plane, a man observes that the angle of elevation of the top of the tower is 16^0. He then moves 150 metres nearer and finds that the angle of elevation of the top is 34^0. Calculate the height of the tower.

27. Kofi observes that the angle of elevation of an aircraft is 35^0 and that 56 seconds later it is 47^0. Assuming that the aircraft is flying at the rate of 270 kilometre an hour in a horizontal line directly over him, calculate its height in kilometres above the ground, and the time, in minutes, which elapses after the first observation before it is overhead.

Chapter Test 1

Take this test as you would take a test in class. After you are done, check your work against the answers in the back of the book

1. Find the value of x. Give your answer correct to 1 decimal place

(a)   (b)   7   (c)

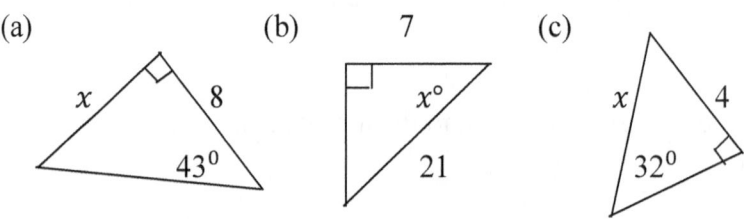

2. A man is 120 m from a building. He finds that the angle of elevation to the top of the building is 23^0. If the man's eye level is 1.6 m above the ground, find the height of the building.

3. Given $\sin A = \frac{3}{5}$, find $2\cos^2 A - 1$

4.

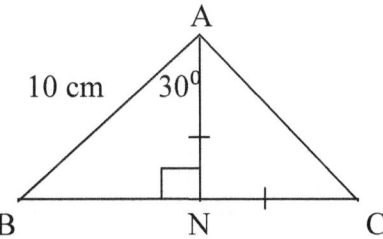

In the diagram above $\angle DCA = 90°$, $\angle DBC = 63°$, $\angle DAC = 35°$ and AB = 12 cm. Calculate correct to one decimal place the length of (a) BC and (b) DC

5. Two boys on opposite sides of a vertical pole are 50 metres apart. Each boy measures the angle of elevation of the top of the pole. If the angles are 36^0 and 47^0, calculate the height of the pole to the nearest tenth.

6. A ship sails directly away from a 75 metre vertical cliff at a constant speed. The angle of depression of the ship viewed at a particular instant from the top of the cliff is 36^0. Then 2 minutes later its angle of depression from the top of the cliff is 25^0. Calculate the speed of the ship in km h^{-1}.

7.

In the diagram $\angle BAN = 30°$, $\angle ANB = 90°$, $AN = NC$ and $AB = 10$ cm. Without using tables or calculator, find the length of AC, giving your answer in the form $k\sqrt{6}$.

8. Two trees are 50 metres apart on a level ground. The height of the taller tree is 65 metres. The angle of depression from the top of the taller tree to the top of the shorter tree is $15°$. Calculate the height of the shorter tree to the nearest tenth of a metre

9.

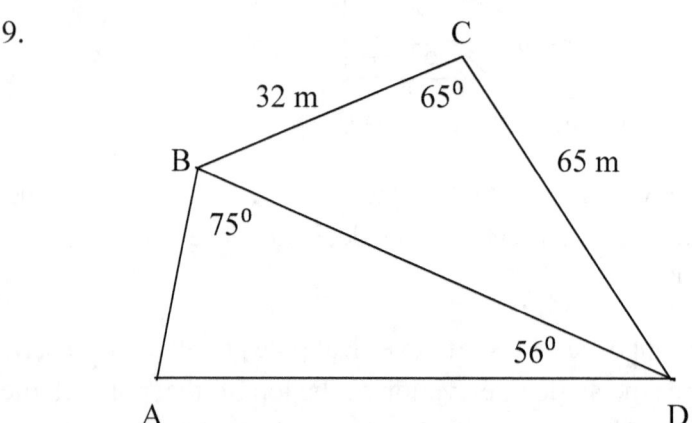

The diagram represents a field in the shape of a quadrilateral ABCD. $BC = 32$ m, $CD = 65$ m, angle $ABD = 75°$, angle $ADB = 56°$ and angle $BCD = 65°$. Calculate, to one decimal place

(a) the length of BD

(b) the perimeter of the whole field ABCD

2

Bearings

Compass Directions

The direction to a point can be started as the number of degrees east or west of North or South. For example, P is 60^0 east of north, written N60^{0}E. The direction of Q is S25^{0}W.

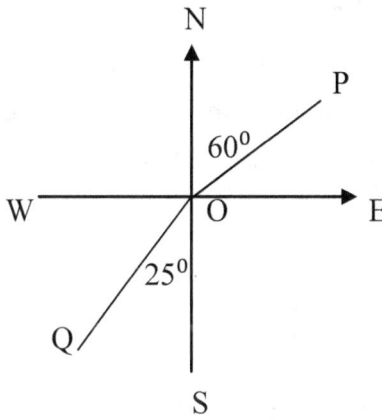

Figure 2.1

The four main points of the compass, called the cardinal points, are North (N), South (S), East (E) and West (W).

The direction half way between the north and east is called North East. Similarly the direction half way between east and south is called South East, the direction half way between west and south is called South West and the direction half way between the west and the north is called North West.

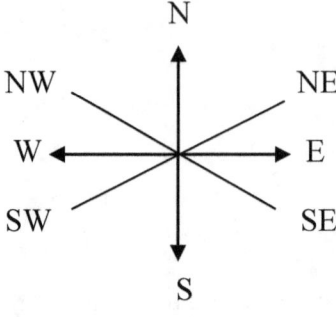

Figure 2.2

The three figure bearings

The bearing to a point is the angle measured clockwise from the North, and is given as a three-digit number. Angles whose measure is one digit or two digits are written with zero(s) in front. Examples are 008^0 and 073^0. The bearing of the North is 000^0.

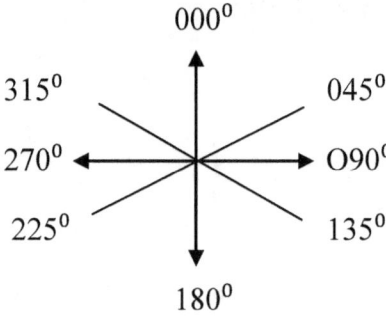

Figure 2.3

2.1 Drawing diagrams to represent bearings

Examples

Draw diagrams to represent the following bearings

(a) Q from P is 036^0 (b) P from O is 217^0

(a) First fixed a point P and then draw a vertical line PN up the page to represent the north line.

Next place the origin of the protractor on P, and align PN with the 0^0 on the protractor. Finally mark where the protractor reads 36^0 Q. Draw a straight line to join P and Q.

Note that the angle is always measured clockwise from the north line.

The diagram shown in Figure 2.4 represents the bearing of Q from P.

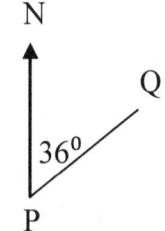

Figure 2.4

(b) Draw the North line ON. Then draw $\angle NOP = 217°$. The diagram may look as shown in Figure 2.5

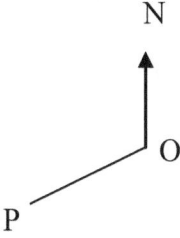

Figure 2.5

Try this 1

Draw a diagram to illustrate the following bearings

(a) B from A is 063^0 (b) Q from P is 112^0

(c) B from A is 215^0 (d) P from O is 345

Measuring Bearings

The diagram below shows two villages at positions A and B. What is the bearing of B from A.?

• B

•

A

First, draw a North line at A (note that A is the starting point). Next, join the points A and B with a straight line. Finally, measure the angle between the North line and line AB, in a clockwise direction. Write down your final bearing. Put zero(s) before your angle if the angle is less than 100^0.

The angle is 65^0, so the bearing of B from A is 065^0.

The diagram will look as shown below.

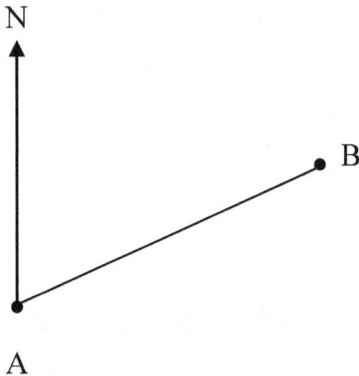

Try this 2

The diagram shows the position of a ship, S and radar, R.

S
●

●

R

Copy the diagram and find the bearing of the ship from the radar.

Back Bearing

If the bearing of B from A is known, then the bearing of A from B is the bearing in the opposite direction, called the back bearing. The two bearings are not equal, and are 180^0 apart as illustrated in the diagram below.

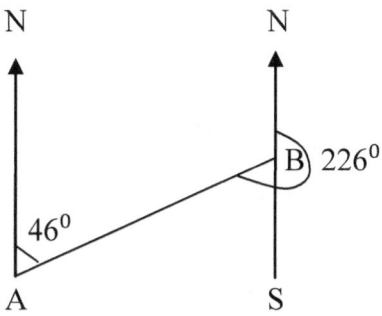

Figure 2.6

In Figure 2.6, angle NAB represents the bearing of B from A, and angle NBA represents the bearing of A from B. Since the two north lines are parallel, angle ABS = 46⁰. Angle NBS is a straight angle, so angle NBA is $(180 + 46)^0$, i.e. 226^0 The bearing of A from B is 226^0

Example

Draw diagrams and use them to find the back bearing of the following bearings

(a) B from A is 037^0 (b) Q from P is 310^0

(a) First, draw a diagram that represents the bearing of B from A.

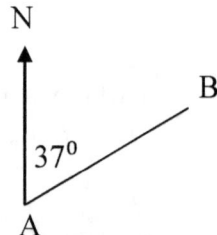

Figure 2.7

Next, draw a north line at B as shown in Figure 2.8

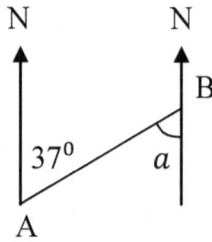

Figure 2.8

The two north lines are parallel so $a = 37°$ (alternate angles). Therefore the bearing of A from B is $(180 +37)^0$. That is 217^0.

(b)

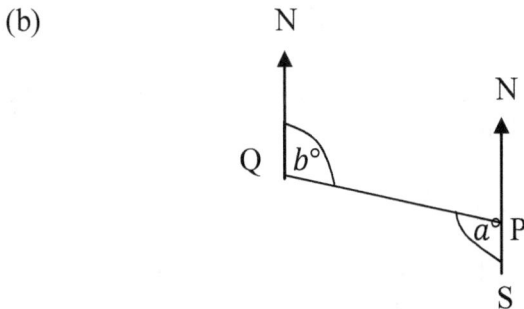

Figure 2.9

Since angle NPS is a straight angle, $a = (310 - 180)°$. That is $a = 130°$. So $b = 130°$, (alternate angles). Therefore, the bearing of P from Q is $130°$.

Try this 3

Draw diagrams and use them to find the back bearing of the following bearings

(a) B from A is 058^0 (b) Q from P is 123^0

(c) A from O is 227^0 (d) P from O is 345^0

You may have noticed that the original bearings and the back bearings are always 180^0 apart. You can find the back bearing without drawing a diagram. If the original bearing is a figure less than 180^0, add 180^0 to it; if the original bearing is a figure greater than 180^0, subtract 180^0 from it.

Exercise 2.1

1. The following are bearings of a point B from a point A, draw diagrams and use them to find the bearings of A from B.

(a) 073^0 (b) 054^0 (c) 108^0 (d) 137^0

(e) 210^0 (f) 265^0 (g) 315^0 (h) 330^0

2. The following are bearings of a point Q from a point P, draw diagrams and use them to find the bearings of P from Q

(a) 035^0 (b) 118^0 (c) 245^0 (d) 327^0

3. The following are bearings of a point B from a point A, without using diagrams find the bearings of A from B

(a) 026^0 (b) 083^0 (c) 127^0 (d) 103^0

(e) 225^0 (f) 283^0 (g) 307^0 (h) 315^0

(i) 000^0 (j) 090^0 (k) 180^0 (l) 270^0

2.2 Calculating Bearings and Distances

Drawing Diagrams

In solving bearing problems, always draw diagrams to help you work out the answers. The following examples illustrate how these diagrams can be drawn.

Examples

The bearing of Q from P is 063^0 and the bearing of R from Q is 145^0. Draw a diagram to represent this information.

First, draw angle NPQ to represent the bearing of Q from P. Then draw angle NQR to represent the bearing of R from Q. Complete the diagram by drawing a line to join P and R. The complete diagram is shown in Figure 2.10

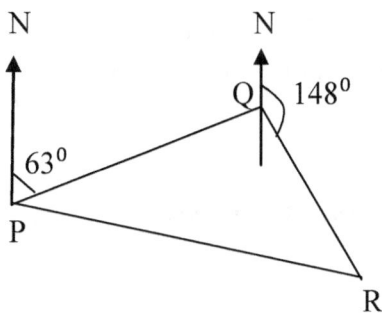

Figure 2.10

Try this 4

The bearing of Q from P is 058^0 and the bearing of R from Q is 125^0. Draw a diagram to represent this information

The bearing of B from A is 126^0 and the bearing of B from C is 233^0. Draw a diagram to represent this information

Draw angle NAB to represent the bearing of B from A. Find the bearing of C from B. Then draw angle NBC to represent the bearing of C from B. Complete the diagram by drawing a line to join A and C .

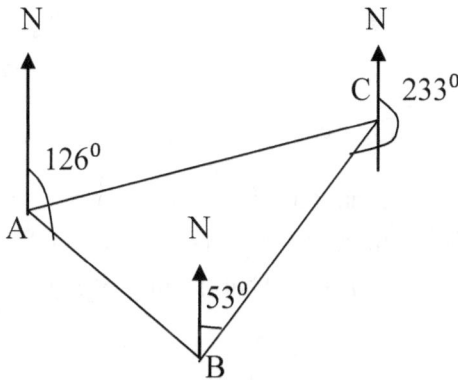

Figure 2.11

Try this 5

The bearing of Q from P is 117^0 and the bearing of Q from R is 256^0. Draw a diagram to represent this information

Finding Bearings and Distances

Examples

The bearing of Q from P is 060^0 and the bearing of R from Q is 150^0. If Q is equidistant from P and R, find the bearing of R from P.

First, draw a diagram that represents the given information (see Figure 2.12)

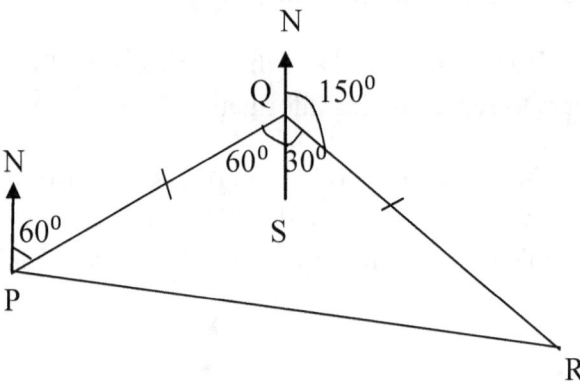

Figure 2.12

Next find an angle in the triangle. Since the two north lines are parallel, angle PQS = 60^0 (alternate angles). Angle NQS is a straight angle, so angle SQR is $(180 - 150)^0$. That is angle SQR = 30^0. Angle PQR is $(60 + 30)^0 = 90^0$

Since P and R are equidistant from Q, triangle PQR is an isosceles triangle. Therefore $\angle QPR = \angle QRP = 45°$. Thus the bearing of R from P is $(60° + 45)° = 105°$.

Try this 6

The bearing of Q from P is 048^0 and the bearing of R from Q is 138^0. If Q is equidistant from P and R, find the bearing of R from P.

A ship sails 5 km on a bearing of 036^0 from P to Q and then 12 km on a bearing of 126^0 from Q to R. Find the distance and bearing of R from P.

The diagram below represents the given information

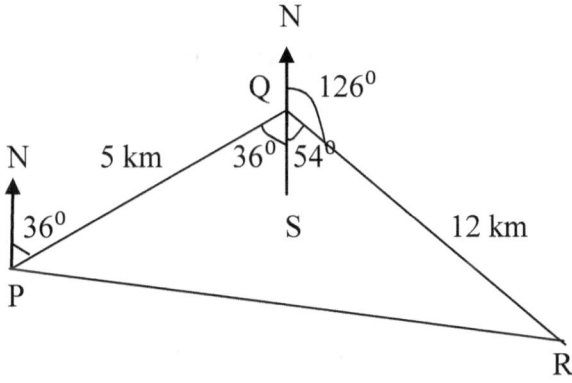

Figure 2.13

You can see from the diagram that $\angle PQR$ is 90^0. Therefore the triangle is a right- angled triangle. Using the Pythagoras' theorem, we have

$$PR^2 = 5^2 + 12^2$$

$$= 25 + 144$$

$$= 169$$

$PR = 13$

Therefore the distance of R from P is 13 km.

$\tan\angle QPR = \dfrac{12}{5}$

$\qquad\quad = 2.4$

$\qquad\quad = 67.38°$

The angle between NP and PR is $36° + 67.38° = 103.38°$

The bearing of R from P is 103^0.

Try this 7

A ship sails 8 km on a bearing of 063^0 from P to Q and then 15 km on a bearing of 153^0 from Q to R. Find the distance and bearing of R from P.

Exercise 2.2

In Problems 1 – 5, draw diagrams to represent the bearings:

1. The bearing of Q from P is 026^0 and the bearing of R from Q is

   218^0

2. From A the bearing of B is 076^0 and from A the bearing of C is

   320^0

3. The bearing of P from Q is 125^0 and the bearing of Q from R is

   030^0

4. The bearing of B from A is 330^0 and the bearing of C from B is 245^0

5. The bearing of Q from P is 340^0 and the bearing of Q from R is 025^0

6. The bearing of Q from P is 300^0 and the bearing of R from Q is 210^0. If Q is equidistant from P and R, find the bearing of R from P.

7. The bearing of Y from X is 035^0 and the bearing of Z from Y is 135^0, find the bearing of Z from X if Y is equidistant from X and Z.

8. The bearing of Q from P is 270^0 and the bearing of R from Q is 050^0. If Q is equidistant from R and P, find the bearing of P from R.

9. From the point X the bearing of Y is 048^0 and the bearing of Z is 300^0. If Y and Z are equidistant from X, find the bearing of Z from Y.

10. The bearing of X from O is 330^0 and the bearing of O from Y is 090^0. If X and Y are equidistant from O, what is the bearing of X from Y?

11. A boat sails from a certain port in the direction N30⁰W. After the boat has sailed 15 kilometres, how far is it west of the port?

12. A cyclist travels 20 kilometres south, then 15 kilometres east. Find the cyclist's bearing from his starting point to the nearest degree.

13. The bearing of Q from P is 060^0 and the bearing of R from Q is 150^0. If P is 30 km from Q and R is 40 km from Q, find the distance and the bearing of R from P.

14. The bearing of Q from P is 120^0 and the bearing of R from Q is 030^0. If P is 12 km from Q and R is 5 km from Q, find the distance and bearing of P from R.

15. Two towns P and Q are situated on a cost and P is due west of Q. From a ship at sea, the bearing of P is 325^0 and the bearing of Q is 055^0. If the distance of the ship from P, and from Q is 8 km and 15 km respectively, calculate the distance PQ.

16. X, Y and Z are three ships at sea. The bearing of Y from X is 060^0 and the bearing of X from Z is 330^0. If XY = 10 km, and XZ = 7 km, calculate correct to three significant figures

(a) YZ

(b) ∠XYZ

(c) the bearing of Z from Y

17. An aircraft flies from A to B on a bearing of 030^0 for $1\frac{1}{2}$ hours at 100 km h^{-1}, and then flies from B to C on a bearing of 120^0 for 1 hour at 80 km h^{-1}. Calculate the distance and bearing of C from A.

18. Two towns P and Q are 35 km apart. Q is on a bearing of 060^0 from P. A town R is on a bearing of 015^0 from P and 330^0 from Q. Calculate, to the nearest kilometre the distance of R from P.

19. From a point on a tower, a man spots a fire 7 km away at a bearing 120^0. The man also spots a village 4 km away on a bearing of 210^0. How far is the fire away from the village? Give your answer correct to the nearest kilometre.

20. A hiker walks from a camp for 8 km on a bearing of 050^0, and then walks a further 12 km on a bearing of 140^0.

(a) How far and on what bearing must he walk to return to the camp?

(b) The hiker walked at a constant speed of $2.5 \, m \, s^{-1}$. If his return journey starts at 3 pm, what time will she arrive at the camp?

21. Two ships R and Q, left a port P, at the same time. Ship R sails at a constant speed of 35 $km \, h^{-1}$ on a bearing of 040^0, and ship Q sails at a constant speed of 40 $km \, h^{-1}$ on a bearing of 310^0. How far apart are the two ships after two hours?

22. A tour boat leaves port and travels 13 km on a bearing of 075^0, and then travels a further 25 km on a bearing of 165^0. The boat then returns directly to the starting point.

(a) Determine the distance to the port and the bearing along which the tour boat must travel.

(b) What is the speed of the boat, if the boat starts the return journey at 10 am and arrives at the starting point at 3 pm?

Review exercise 2

In Problems 1 – 2, draw diagrams to represent the bearings

1. The bearing of Q from P is 035^0 and the bearing of R from Q is 228^0

2. The bearing of B from A is 320^0 and the bearing of C from B is 213^0

3. Use diagrams to find the back bearing of the following bearings:

(a) B from A is 086^0 (b) Q from P is 218^0

(c) P from O is 153^0 (d) Q from P is 314^0

4. Without using diagrams, find the back bearing of

(a) 043^0 (b) 320^0 (c) 175^0 (d) 259^0

5. The bearing of Q from P is 330^0 and the bearing of R from Q is 210^0. If Q is equidistant from P and R, find the bearing of R from P.

6. The bearing of Y from X is 030^0 and the bearing of Z from Y is 120^0. Find the bearing of Z from X if Y is equidistant from X and Z.

7. The bearing of Y from X is 042^0 and the bearing of Y from Z is 330^0, find the bearing of Z from X if Y is equidistant from X and Z.

8. The bearing of Q from P is 120^0 and the bearing of R from Q is 030^0. If P is 12 km from Q and R is 5 km from Q, find the distance and bearing of R from P.

9. The bearing of Q from P is 130^0 and the bearing of R from Q is 040^0. If P is 15 km from Q and R is 8 km from Q, find the distance and bearing of P from R.

10. The bearing of Q from P is 036^0 and the bearing of Q from R is 306^0. If P is 40 km from Q and R is 30 km from Q, find the distance and the bearing of P from R.

Chapter Test 2

Take this test as you would take a test in class. After you are done, check your work against the answers in the back of the book.

1. Use diagrams to find the back bearing of the following bearings:

(a) B from A is 108^0 (b) Q from P is 247^0

2. Without using diagrams, find the back bearing of

(a) 045^0 (b) 308^0

3. Copy the diagram below and find the bearing of Q from P.

Q
•

•

P

4. The bearing of Q from P is 315^0 and the bearing of R from Q is 205^0. If Q is equidistant from P and R, find the bearing of R from P.

5. The bearing of Y from X is 025^0 and the bearing of Z from Y is 145^0, find the bearing of Z from X if Y is equidistant from X and Z.

6. A ship sails 100 kilometres from A on a bearing of 070^0 to B. It then sails 120 kilometres on a bearing of 160^0 to C. Find correct to the nearest unit , the distance and bearing of C from A.

7. A man drives 13 kilometres from A on a bearing of 040^0 to B and then drives 17 kilometres to C on a bearing of 130^0. Find correct to the nearest unit, the distance and bearing of A from C.

8. A, B and C are three ships at sea. The bearing of A from B is 045^0. The bearing of C from A is 135^0. If AB = 15 km and AC = 8 km, what is the distance and bearing of B from C.

3

Vectors

A vector describe a movement from one point to another

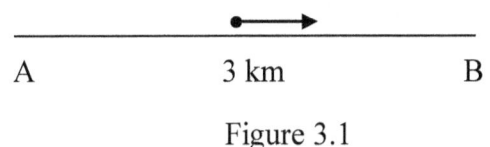

A   3 km   B

Figure 3.1

A car at A travels 3 kilometres east to B, represented by the diagram shown in Figure 3.1. We have described the car's journey by stating how far it travelled and in which direction it travelled. Certain quantities can only be described fully by stating a magnitude (size) and a direction. Such quantities are called vectors. Examples of vector quantities are displacement, velocity and forces. Other quantities, such as area, time and temperature can be represented by a single real number. Such quantities have only magnitude, and are called scalar quantities.

Vectors are represented diagrammatically as directed line segments, that is line segments with specific directions as shown in Figure 3.2

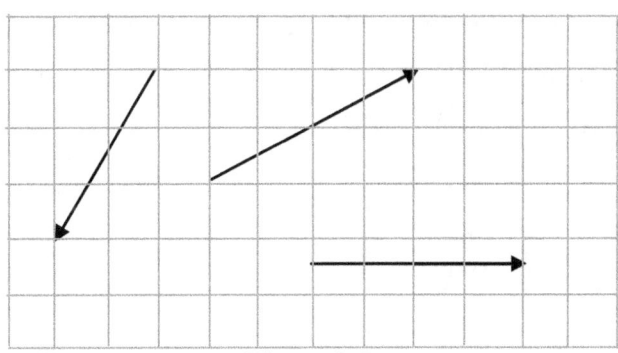

Figure 3.2

The direction of the arrow indicates the direction of the vector, and the length of the line segment represents the magnitude of the vector.

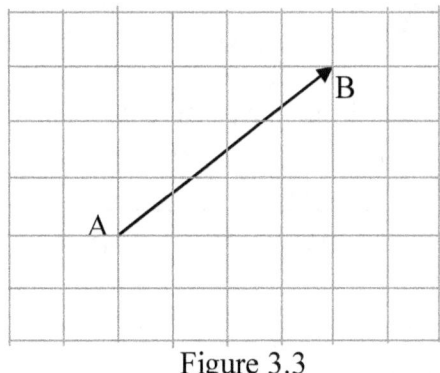

Figure 3.3

Figure 3.3 shows a vector with initial point at A and terminal point at B. The vector is written as $\overrightarrow{AB}$. In print vectors are written in bold type such as **AB**

Vectors can be represented by single letters such as **a** . In handwriting you should indicate a vector by arrow or by a line under the letter, for example $\overrightarrow{AB}$ or $\underline{a}$.

Equal Vectors

Two vectors are equal if they have the same magnitude and direction regardless of the location of their initial point. If two vectors $\overrightarrow{AB}$ and $\overrightarrow{PQ}$ are equal, then we write $\overrightarrow{AB} = \overrightarrow{PQ}$.

Unit Vectors

A unit vector has a magnitude of 1.

Magnitude- Bearing Form of a Vector

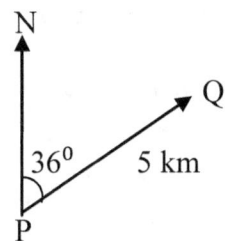

Figure 3.4

Figure 3.4 illustrates the vector from P to Q, written as $\overrightarrow{PQ}$. The magnitude of this vector is 5 km, and its direction is 036^0. This can be written in brief as $\overrightarrow{PQ} = (5\ km, 036°)$. This representation of a vector is called the magnitude- bearing form. The length of the directed line segment $\overrightarrow{PQ}$ is the magnitude of the vector, and is denoted by $\left|\overrightarrow{PQ}\right|$ or PQ.

3.1 Component Form of a Vector

A vector can be represented by its movement in both the x direction and the y direction. The movement in the x direction is called the horizontal component, and the movement in the y direction is called the vertical component. This is represented by a column vector.

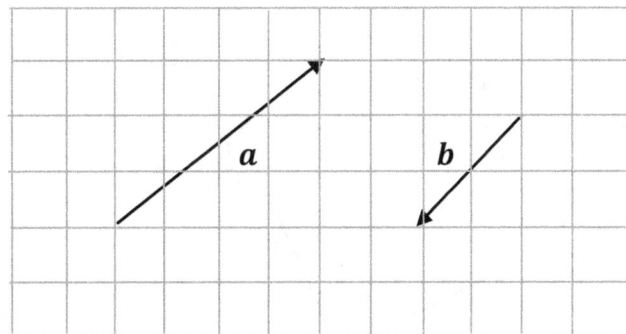

Figure 3.5

In Figure 3.5 vector **a** moves 4 units in the x direction, and then 3 units in the y direction. That is $\boldsymbol{a} = \begin{pmatrix} 4 \\ 3 \end{pmatrix}$. The vector **b** is $\boldsymbol{b} = \begin{pmatrix} -2 \\ -2 \end{pmatrix}$.

You can see from Figure 3.5 that in the x direction, movement to the right is positive and movement to the left is negative. In the y direction, movement up is positive and movement down is negative.

Try this 1

Write the vectors in the diagrams as column vectors

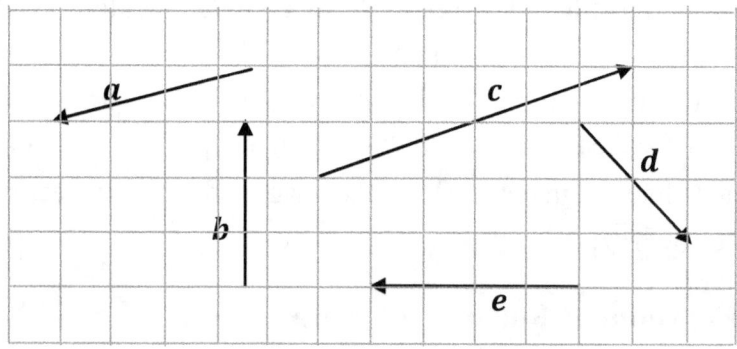

Exercise 3.1(a)

1

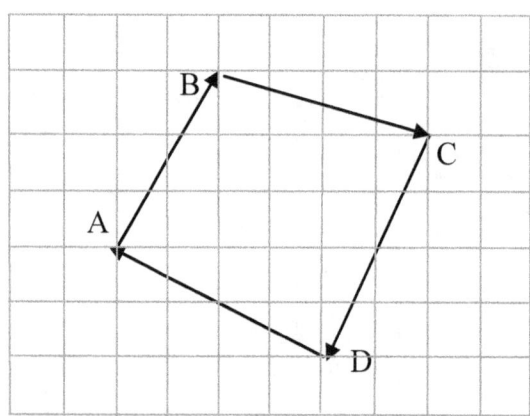

Write the following vectors as column vectors:

(a) $\overrightarrow{AB}$ (b) $\overrightarrow{BC}$ (c) $\overrightarrow{CD}$ (d) $\overrightarrow{DA}$

2. On a square grid draw diagrams to represent the following vectors

(a) $a = \begin{pmatrix} 4 \\ -3 \end{pmatrix}$ (b) $b = \begin{pmatrix} -2 \\ 1 \end{pmatrix}$ (c) $c = \begin{pmatrix} -3 \\ 0 \end{pmatrix}$ (d) $d = \begin{pmatrix} -2 \\ -2 \end{pmatrix}$

(e) $e = \begin{pmatrix} 0 \\ 2 \end{pmatrix}$ (f) $f = \begin{pmatrix} 3 \\ -1 \end{pmatrix}$ (g) $g = \begin{pmatrix} 2 \\ 0 \end{pmatrix}$ (h) $h = \begin{pmatrix} 0 \\ -1 \end{pmatrix}$

Negative Vectors

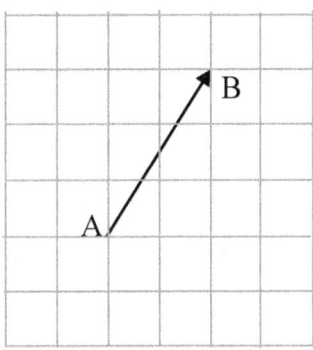

Figure 3.6

The diagram in Figure 3.6 represents vector $\overrightarrow{AB}$. The vector in the opposite direction, which is vector $\overrightarrow{BA}$, is the negative vector of $\overrightarrow{AB}$. So $\overrightarrow{BA} = -\overrightarrow{AB}$. Notice that $\overrightarrow{BA} = \begin{pmatrix} -2 \\ -3 \end{pmatrix}$.

In general, the negative of a vector **a,** denoted $-$ **a** is a vector which has the same magnitude as **a** but has a direction opposite to **a.** If $a = \begin{pmatrix} a_1 \\ a_2 \end{pmatrix}$, then $- a = \begin{pmatrix} -a_1 \\ -a_2 \end{pmatrix}$

Example

Find the negative vector of each of the following vectors

(a) $a = \begin{pmatrix} -3 \\ 4 \end{pmatrix}$ (b) $x = \begin{pmatrix} 5 \\ -2 \end{pmatrix}$

(a) $- a = \begin{pmatrix} 3 \\ -4 \end{pmatrix}$

(b) $- x = \begin{pmatrix} -5 \\ 2 \end{pmatrix}$

Try this 2

Find the negative vector of each of the following vectors

(a) $a = \begin{pmatrix} -7 \\ -3 \end{pmatrix}$ (b) $x = \begin{pmatrix} 0 \\ 4 \end{pmatrix}$

Exercise 3.1(b)

Find the negative vector of each of the following vectors

1. $\begin{pmatrix} 3 \\ 4 \end{pmatrix}$ 2. $\begin{pmatrix} 0 \\ 2 \end{pmatrix}$ 3. $\begin{pmatrix} -5 \\ -4 \end{pmatrix}$ 4. $\begin{pmatrix} -1 \\ 3 \end{pmatrix}$ 5. $\begin{pmatrix} 6 \\ -2 \end{pmatrix}$ 6. $\begin{pmatrix} -7 \\ 0 \end{pmatrix}$

3.2 Operations with Vectors

Addition of Vectors

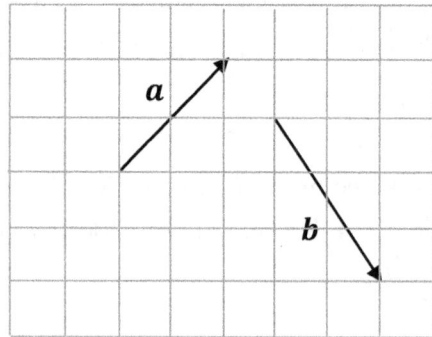

Figure 3.7

You can add the vectors **a** and **b** shown in Figure 3.7 by putting the initial point of **b** and the terminal point of **a** together, as shown in Figure 3.8.

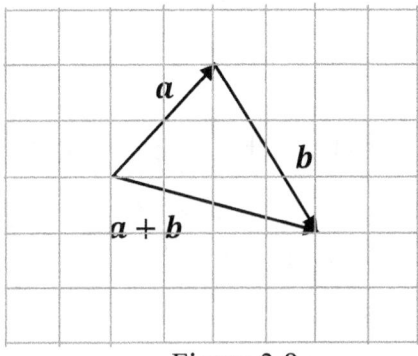

Figure 3.8

The sum of vector **a** and **b** is the vector from the initial point of **a** to the terminal of **b**. The sum (**a** + **b**), of the vector is called the resultant.

You can see from Figure 3.8 that $a + b = \begin{pmatrix} 4 \\ -1 \end{pmatrix}$.

Notice that $\begin{pmatrix} 2 \\ 2 \end{pmatrix} + \begin{pmatrix} 2 \\ -3 \end{pmatrix} = \begin{pmatrix} 4 \\ -1 \end{pmatrix}$

In general, if $a = \begin{pmatrix} a_1 \\ a_2 \end{pmatrix}$ and $b = \begin{pmatrix} b_1 \\ b_2 \end{pmatrix}$ then $a + b = \begin{pmatrix} a_1 + b_1 \\ a_2 + b_2 \end{pmatrix}$

Example

Work out

$$\begin{pmatrix} 3 \\ -5 \end{pmatrix} + \begin{pmatrix} -4 \\ 7 \end{pmatrix}$$

$$\begin{pmatrix} 3 \\ -5 \end{pmatrix} + \begin{pmatrix} -4 \\ 7 \end{pmatrix} = \begin{pmatrix} -1 \\ 2 \end{pmatrix}$$

Try this 3

Find $\begin{pmatrix} 2 \\ 1 \end{pmatrix} + \begin{pmatrix} 1 \\ -3 \end{pmatrix}$

Exercise 3.2(a)

1. Work out:

(a) $\begin{pmatrix} 3 \\ 2 \end{pmatrix} + \begin{pmatrix} 1 \\ 3 \end{pmatrix}$
(b) $\begin{pmatrix} -2 \\ 1 \end{pmatrix} + \begin{pmatrix} 4 \\ -5 \end{pmatrix}$
(c) $\begin{pmatrix} 6 \\ -7 \end{pmatrix} + \begin{pmatrix} -1 \\ 5 \end{pmatrix}$

(d) $\begin{pmatrix} -8 \\ -6 \end{pmatrix} + \begin{pmatrix} 10 \\ 4 \end{pmatrix}$
(e) $\begin{pmatrix} 0 \\ -3 \end{pmatrix} + \begin{pmatrix} -2 \\ 5 \end{pmatrix}$
(f) $\begin{pmatrix} -2 \\ -3 \end{pmatrix} + \begin{pmatrix} -3 \\ 5 \end{pmatrix}$

2. If $a = \begin{pmatrix} 5 \\ -3 \end{pmatrix}$ and $b = \begin{pmatrix} -2 \\ 7 \end{pmatrix}$, find

(a) $a + b$ (b) $b + a$

What property is shown by the results in (a) and (b)

3. If $a = \begin{pmatrix} 2 \\ -4 \end{pmatrix}$, $b = \begin{pmatrix} -3 \\ 2 \end{pmatrix}$ and $c = \begin{pmatrix} 4 \\ 3 \end{pmatrix}$, find

(a) $(a + b) + c$ (b) $a + (b + c)$

What property is shown by the results in (a) and (b)

Subtraction of Vectors

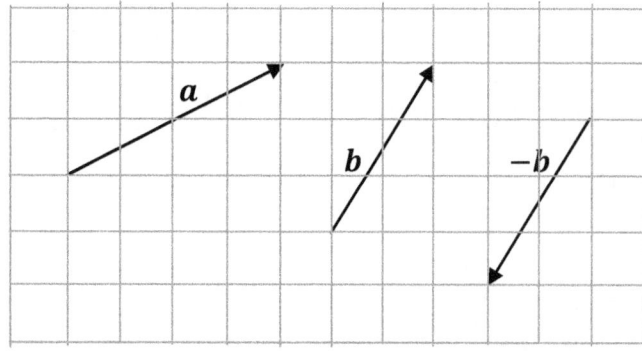

Figure 3.9

You can subtract **b** from **a** by adding the negative vector of **b** i.e. − **b** to **a** as shown in Figure 3.10. That is $a + (-b) = a - b$.

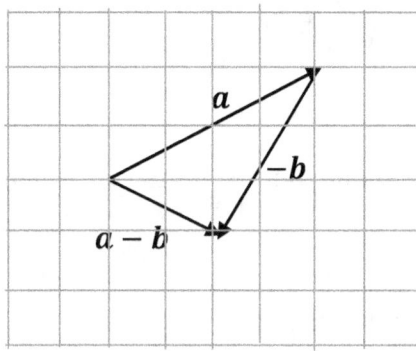

Figure 3.10

You can see from Figure 3.10 that $\binom{4}{2} - \binom{2}{3} = \binom{2}{-1}$

In general if $a = \binom{a_1}{a_2}$ and $b = \binom{b_1}{b_2}$ then $a - b = \binom{a_1 - b_1}{a_2 - b_2}$

Example

Find $\binom{8}{-3} - \binom{5}{-7}$

$\binom{8}{-3} - \binom{5}{-7} = \binom{3}{4}$

Try this 4

Find $\binom{-7}{2} - \binom{-9}{5}$

Exercise 3.2(b)

1. Find

(a) $\binom{6}{8} - \binom{5}{9}$ (b) $\binom{0}{-2} - \binom{3}{1}$ (c) $\binom{-8}{-2} - \binom{-5}{-4}$

(d) $\begin{pmatrix} 3 \\ 5 \end{pmatrix} - \begin{pmatrix} 1 \\ 8 \end{pmatrix}$ (e) $\begin{pmatrix} -2 \\ 6 \end{pmatrix} - \begin{pmatrix} -2 \\ 4 \end{pmatrix}$ (f) $\begin{pmatrix} -3 \\ 2 \end{pmatrix} - \begin{pmatrix} 0 \\ -4 \end{pmatrix}$

2. If $a = \begin{pmatrix} -3 \\ 5 \end{pmatrix}$ and $b = \begin{pmatrix} 2 \\ -1 \end{pmatrix}$, find

   (a) $a - b$ (b) $b - a$

   What do you notice about your results to (a) and (b)

Multiplication by a Scalar

The product of a vector **x** and a scalar (a real number) k is the vector that is $|k|$ times as long as **x**. If k is positive $k\mathbf{x}$ has the same direction as **x**, and if k is negative, $k\mathbf{a}$ has opposite direction of **x**.

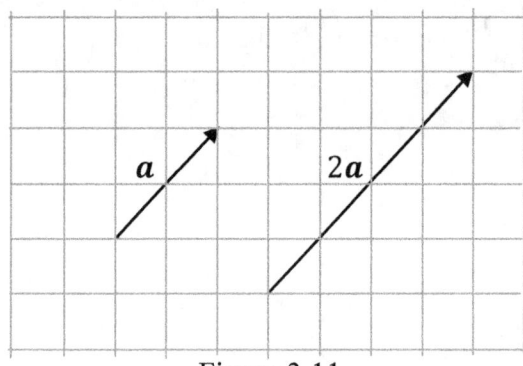

Figure 3.11

Notice that $2\mathbf{a} = 2\begin{pmatrix} 2 \\ 2 \end{pmatrix} = \begin{pmatrix} 4 \\ 4 \end{pmatrix}$. Notice that each component is multiplied by 2.

In general, given $a = \begin{pmatrix} a_1 \\ a_2 \end{pmatrix}$ and a scalar k, then $k\mathbf{a} = \begin{pmatrix} ka_1 \\ ka_2 \end{pmatrix}$

Examples

Find $-3\begin{pmatrix} 2 \\ -1 \end{pmatrix}$

$$-3\begin{pmatrix} 2 \\ -1 \end{pmatrix} = \begin{pmatrix} -6 \\ 3 \end{pmatrix}$$

Try this 5

Find $\frac{1}{2}\begin{pmatrix} -4 \\ 6 \end{pmatrix}$

Given $\boldsymbol{a} = \begin{pmatrix} 3 \\ -2 \end{pmatrix}$ and $\boldsymbol{b} = \begin{pmatrix} 1 \\ 4 \end{pmatrix}$, find $2\boldsymbol{a} + 3\boldsymbol{b}$

Begin by substituting the vectors

$$2\boldsymbol{a} + 3\boldsymbol{b} = 2\begin{pmatrix} 3 \\ -2 \end{pmatrix} + 3\begin{pmatrix} 1 \\ 4 \end{pmatrix}$$

$$= \begin{pmatrix} 6 \\ -4 \end{pmatrix} + \begin{pmatrix} 3 \\ 12 \end{pmatrix} \quad \text{Multiplying the vectors by the scalar}$$

$$= \begin{pmatrix} 9 \\ 8 \end{pmatrix} \quad\quad\quad \text{Adding the vectors}$$

Try this 6

Given $\boldsymbol{a} = \begin{pmatrix} -2 \\ 3 \end{pmatrix}$ and $\boldsymbol{b} = \begin{pmatrix} -4 \\ -1 \end{pmatrix}$, find $3\boldsymbol{a} - 2\boldsymbol{b}$

Exercise 3.2(c)

1. Find:

(a) $-2\begin{pmatrix} -3 \\ 2 \end{pmatrix}$ (b) $5\begin{pmatrix} 1 \\ 0 \end{pmatrix}$ (c) $-4\begin{pmatrix} -1 \\ -2 \end{pmatrix}$ (d) $3\begin{pmatrix} 3 \\ 1 \end{pmatrix}$

(e) $\frac{2}{3}\begin{pmatrix} 6 \\ 9 \end{pmatrix}$ (f) $-\frac{1}{2}\begin{pmatrix} -4 \\ 8 \end{pmatrix}$ (g) $\frac{3}{2}\begin{pmatrix} 4 \\ -6 \end{pmatrix}$ (h) $-\frac{3}{4}\begin{pmatrix} -8 \\ 12 \end{pmatrix}$

2. If $a = \begin{pmatrix} 3 \\ 1 \end{pmatrix}$, $b = \begin{pmatrix} -2 \\ 4 \end{pmatrix}$ and $c = \begin{pmatrix} -5 \\ -3 \end{pmatrix}$, find

(a) $2a + b$ (b) $5a + 4c$ (c) $2b - 3a$ (d) $2c - \frac{3}{2}b$

3. Given $x = \begin{pmatrix} 6 \\ 8 \end{pmatrix}$, $y = \begin{pmatrix} a \\ b \end{pmatrix}$ and $z = \begin{pmatrix} 12 \\ -5 \end{pmatrix}$, find y if $\frac{1}{2}x + 3y = z$

4. Given $x = \begin{pmatrix} 2 \\ 3 \end{pmatrix}$, $y = \begin{pmatrix} 3 \\ -1 \end{pmatrix}$ and $z = \begin{pmatrix} 12 \\ 7 \end{pmatrix}$, find a and b if $ax + by = z$

5. Given $x = \begin{pmatrix} a \\ 3 \end{pmatrix}$, $y = \begin{pmatrix} b \\ a \end{pmatrix}$ and $z = \begin{pmatrix} 5 \\ 0 \end{pmatrix}$, find the values of a and b if $2x + 3y = z$

3.3 Magnitude of Vectors

Recall that the magnitude of a vector is represented by the length of the vector, and the magnitude of $\overrightarrow{AB}$ or $\mathbf{a}$ is denoted by $|\overrightarrow{AB}|$ or $|a|$.

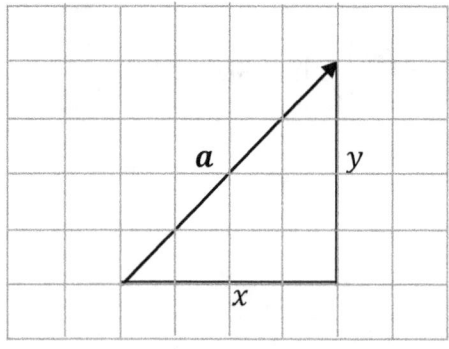

Figure 3.12

Figure 3.12 shows the vector $a = \begin{pmatrix} x \\ y \end{pmatrix}$. Using the Pythagoras' theorem, we have

$$|a| = \sqrt{x^2 + y^2}$$

In general, the magnitude of a vector $\begin{pmatrix} x \\ y \end{pmatrix}$ is $\sqrt{x^2 + y^2}$.

Example

Calculate the magnitude of the vector $\overrightarrow{PQ} = \begin{pmatrix} 5 \\ 12 \end{pmatrix}$

$|\overrightarrow{PQ}| = \sqrt{5^2 + 12^2}$

$\qquad = \sqrt{25 + 144}$

$\qquad = \sqrt{169}$

$\qquad = 13$

The magnitude of $\overrightarrow{PQ}$ is 13 units.

Try this 7

Calculate the magnitude of the vector $a = \begin{pmatrix} -8 \\ 6 \end{pmatrix}$

Exercise 3.3

Find the magnitude of each of the following vectors

1. $\begin{pmatrix} 3 \\ -4 \end{pmatrix}$ 2. $\begin{pmatrix} 8 \\ 15 \end{pmatrix}$ 3. $\begin{pmatrix} -12 \\ 9 \end{pmatrix}$ 4. $\begin{pmatrix} -5 \\ -8 \end{pmatrix}$ 5. $\begin{pmatrix} 6 \\ 4 \end{pmatrix}$ 6. $\begin{pmatrix} -5 \\ 12 \end{pmatrix}$

3.4 Angles between a vector and the x-axis

Consider the vector $a = \begin{pmatrix} x \\ y \end{pmatrix}$, where x represent the displacement in x direction and y represent the displacement in the y direction as shown in Figure 3.13.

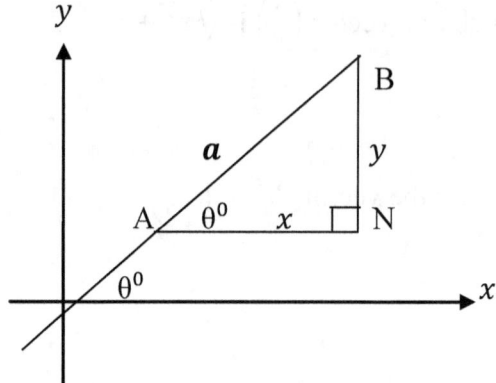

Figure 3.13

Let the angle between the vector **a** and the x- axis be θ. Since AN is parallel to the x-axis $\angle NAB = \theta^{\circ}$ (corresponding angles). From triangle ABN

$$\tan \theta^{\circ} = \frac{y}{x}$$

Thus, $\theta = tan^{-1}\left(\frac{y}{x}\right)$

Hence, the angle between the vector and the x- axis is $tan^{-1}\left(\frac{y}{x}\right)$

Example

Find the angle between the vector $\overrightarrow{PQ} = \begin{pmatrix} 8 \\ 15 \end{pmatrix}$ and the x- axis to the nearest degree

Let θ represent the angle between the vector $\overrightarrow{PQ}$ and the x- axis

$$\tan \vartheta^{\circ} = \frac{15}{8}$$

$$= 1.875$$

$$\theta = 61.92^{\circ}$$

$= 62°$

Try this 8

Find to the nearest degree the angle between the vector $a = \begin{pmatrix} 5 \\ 12 \end{pmatrix}$ and the x-axis.

Exercise 3.4

Find to the nearest degree the acute angle between each of the following vectors and the x- axis:

1. $\begin{pmatrix} 3 \\ 4 \end{pmatrix}$ 2. $\begin{pmatrix} 6 \\ 15 \end{pmatrix}$ 3. $\begin{pmatrix} -4 \\ 5 \end{pmatrix}$ 4. $\begin{pmatrix} 8 \\ -10 \end{pmatrix}$ 5. $\begin{pmatrix} -12 \\ -9 \end{pmatrix}$

3.5 Changing a vector from one form to another form

Charging a vector from a magnitude- bearing form to a column vector

Example

Express $\overrightarrow{OA} = (10\ units, 120°)$ as a column vector

Draw a diagram to represent the vector.

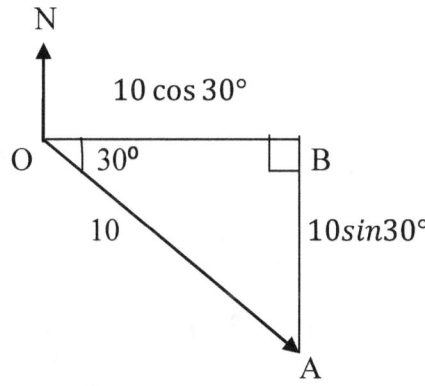

Figure 3.14

Figure 3.14   shows   the   vector $\overrightarrow{OA} = (10\ units, 120°)$.   From trigonometry $|OB| = 10\cos 30°$ and $|AB| = 10\sin 30°$. Thus,

$$\overrightarrow{OA} = \begin{pmatrix} 10\cos 30° \\ -10\sin 30° \end{pmatrix}$$

$$= \begin{pmatrix} 10 \times 0.866 \\ -10 \times 0.5 \end{pmatrix}$$

$$= \begin{pmatrix} 8.66 \\ -5 \end{pmatrix}$$

Try this 9

Express $\overrightarrow{AB} = (50\ units,\ 030°)$ as a column vector

Changing a column vector to the magnitude- bearing form

Example

Change the vector $\overrightarrow{OA} = \begin{pmatrix} -8 \\ -6 \end{pmatrix}$ to the magnitude - bearing form

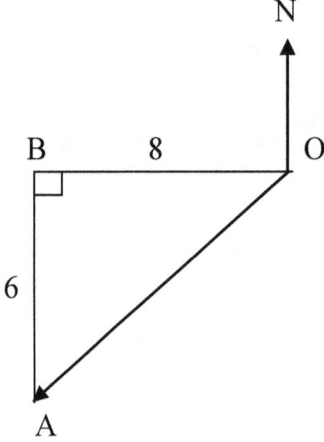

Figure 3.15

Triangle OAB is a right - angled triangle. By Pythagoras' theorem

$$OA^2 = OB^2 + AB^2$$

$$= (-8)^2 + (-6)^2$$

$$= 64 + 36$$

$$= 100$$

$$OA = 10$$

The magnitude of the vector $\overrightarrow{OA}$ is 10 units

$$\tan \angle AOB = \frac{6}{8}$$

$$= 0.75$$

$$\angle AOB = 37°$$

The direction of the vector is $(270 - 37)° = 233°$

Therefore, $\overrightarrow{OA} = (10 \; units, 233°)$

Try this 10

Change $a = \begin{pmatrix} -3 \\ 4 \end{pmatrix}$ to the magnitude - bearing form

Exercise 3.5

1. Express each of the following vectors as a column vector

(a) $(50 \; units, 046°)$ (b) $(100 \; units, 135°)$

(c) $(5 \; units, 236°)$ (d) $(120 \; units, 330°)$

(e) $(12 \; units, 210°)$ (f) $(15 \; units, 315°)$

2. Find the magnitude - bearing form of each of the following vectors

(a) $\begin{pmatrix} 5 \\ 12 \end{pmatrix}$ (b)$\begin{pmatrix} -6 \\ 8 \end{pmatrix}$ (c) $\begin{pmatrix} 5 \\ -9 \end{pmatrix}$ (d) $\begin{pmatrix} -11 \\ -16 \end{pmatrix}$ (e) $\begin{pmatrix} 8 \\ -15 \end{pmatrix}$

3.6 Solving Vectors by Calculation

In solving vector problems, diagrams can be useful. It helps you see what calculations to do.

Example

Find $\overrightarrow{AC}$ if $\overrightarrow{AB} = \begin{pmatrix} -2 \\ 3 \end{pmatrix}$ and $\overrightarrow{BC} = \begin{pmatrix} 6 \\ -2 \end{pmatrix}$.

Begin by drawing a diagram

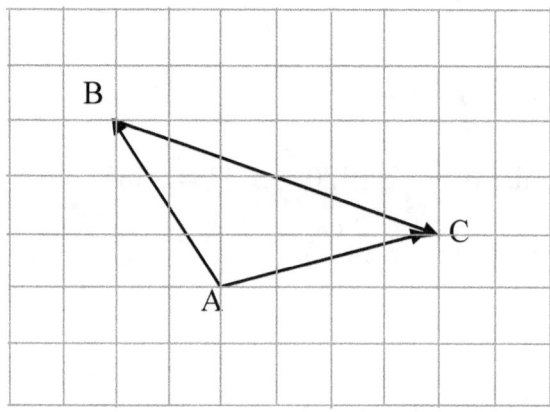

You can see from the diagram that

$\overrightarrow{AC} = \overrightarrow{AB} + \overrightarrow{BC}$

$\quad = \begin{pmatrix} -2 \\ 3 \end{pmatrix} + \begin{pmatrix} 6 \\ -2 \end{pmatrix}$

$\quad = \begin{pmatrix} 4 \\ 1 \end{pmatrix}$

Try this 11

If $\overrightarrow{PQ} = \begin{pmatrix} 2 \\ 1 \end{pmatrix}$ and $\overrightarrow{PR} = \begin{pmatrix} -4 \\ 5 \end{pmatrix}$, find $\overrightarrow{QR}$

Exercise 3.6

1. Find $\overrightarrow{ZY}$ if $\overrightarrow{XY} = \begin{pmatrix} -3 \\ 2 \end{pmatrix}$ and $\overrightarrow{ZX} = \begin{pmatrix} 3 \\ 4 \end{pmatrix}$

2. If $\overrightarrow{AB} = \begin{pmatrix} 4 \\ 5 \end{pmatrix}$ and $\overrightarrow{BC} = \begin{pmatrix} 2 \\ -4 \end{pmatrix}$, find $\overrightarrow{AC}$

3. Find $\overrightarrow{AC}$ if $\overrightarrow{AB} = \begin{pmatrix} 5 \\ -6 \end{pmatrix}$ and $\overrightarrow{CB} = \begin{pmatrix} -7 \\ 3 \end{pmatrix}$

4. If $\overrightarrow{AB} = \begin{pmatrix} 2 \\ 3 \end{pmatrix}$, $\overrightarrow{BC} = \begin{pmatrix} 3 \\ -2 \end{pmatrix}$ and $\overrightarrow{CD} = \begin{pmatrix} -2 \\ -2 \end{pmatrix}$, find $\overrightarrow{AC}$ and $\overrightarrow{AD}$

5. If $\overrightarrow{QR} = \begin{pmatrix} -2 \\ -3 \end{pmatrix}$, $\overrightarrow{QS} = \begin{pmatrix} -3 \\ 2 \end{pmatrix}$ and $\overrightarrow{ST} = \begin{pmatrix} -3 \\ 1 \end{pmatrix}$, calculate $\overrightarrow{QT}$ and $\overrightarrow{RS}$

3.7 Coordinates and Vectors

A vector drawn with its initial point located at the origin is called the position vector. For example a vector that has its initial point at O and the terminal point at A is called the position vector of A, denoted by $\overrightarrow{OA}$.

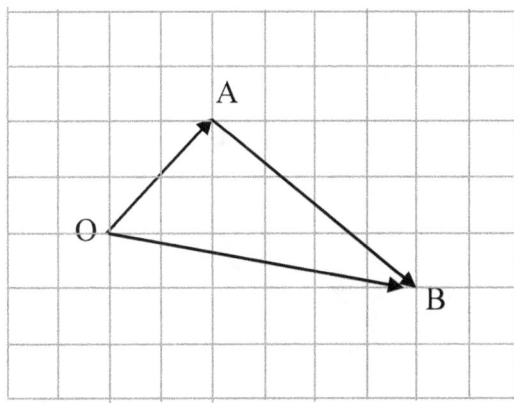

Figure 3.16

In Figure 3.16, O is the origin. The position vector of A is $\overrightarrow{OA} = \binom{2}{2}$.
You may have noticed that the coordinate of A is (2, 2). The position of
B is $\overrightarrow{OB} = \binom{6}{-1}$

You can express vector $\overrightarrow{AB}$ in terms of the position vector of A and
B.

$$\overrightarrow{AB} = \overrightarrow{AO} + \overrightarrow{OB}$$

Since vector addition is commutative, and $\overrightarrow{AO} = -\overrightarrow{OA}$, we have

$$\overrightarrow{AB} = \overrightarrow{OB} - \overrightarrow{OA}$$

$$= \binom{6}{-1} - \binom{2}{2}$$

$$= \binom{4}{-3}$$

In general, given the coordinates of any two points $A(a_1, a_2)$ and
$B(b_1, b_2)$ the vector $\overrightarrow{AB}$ can be expressed as

$$\overrightarrow{AB} = \overrightarrow{OB} - \overrightarrow{OA} = \binom{b_1 - a_1}{b_2 - a_2}.$$

Examples

The coordinates of P and Q are (2, -3) and (- 1, 2) respectively. Find
the vector $\overrightarrow{PQ}$.

$$\overrightarrow{PQ} = \overrightarrow{OQ} - \overrightarrow{OP}$$

$$= \binom{-1}{2} - \binom{2}{-3}$$

$$= \binom{-3}{5}$$

Try this 12

The coordinates of A and B are (- 3,2) and (5, - 4) respectively. Find the vector $\overrightarrow{AB}$.

Given $A(x, y)$, $B(4, -2)$ and $\overrightarrow{AB} = \begin{pmatrix} -2 \\ 1 \end{pmatrix}$, find the values of x and y.

$$\overrightarrow{AB} = \overrightarrow{OB} - \overrightarrow{OA}$$

$$\begin{pmatrix} -2 \\ 1 \end{pmatrix} = \begin{pmatrix} 4 \\ -2 \end{pmatrix} - \begin{pmatrix} x \\ y \end{pmatrix}$$

$$= \begin{pmatrix} 4 - x \\ -2 - y \end{pmatrix}$$

Since the corresponding components of the two vectors are equal, we have

$$-2 = 4 - x$$

$$x = 6$$

and $1 = -2 - y$

$$y = -3$$

Try this 13

Given $P(x, y)$, $Q(-3,2)$ and $\overrightarrow{PQ} = \begin{pmatrix} 2 \\ 1 \end{pmatrix}$, find the values of x and y.

Exercise 3.7

1. If $A(4,3)$ and $B(6,4)$, calculate the vectors $\overrightarrow{AB}$ and $\overrightarrow{BA}$

2. Given that $P(-3,2)$ and $Q(2,-1)$ calculate the vector $\overrightarrow{PQ}$

3. If $P(-5,4), Q(-4,6)$ and $R(3,2)$, calculate the vectors $\overrightarrow{PQ}, \overrightarrow{PR}$ and $\overrightarrow{QR}$.

4. Given $A(-5,4)$ and $\overrightarrow{AB} = \begin{pmatrix} 8 \\ -6 \end{pmatrix}$, find the coordinates of B.

5. If $P(-3,4)$ and $\overrightarrow{PQ} = \begin{pmatrix} 3 \\ 3 \end{pmatrix}$, find the coordinates of Q.

6. If $Q(-3,2)$ and $\overrightarrow{PQ} = \begin{pmatrix} -7 \\ 5 \end{pmatrix}$, find the coordinates of P.

3.8 Using Vectors in Geometry

Parallel Vectors

Two vectors having the same or opposite direction are called parallel vectors. If two vectors **a** and **b** are parallel, written **a**//**b**, then **a** is a scalar multiple of **b**, that is $\mathbf{a} = k\mathbf{b}$, where k is a scalar

Example

Determine whether or not the vectors $\mathbf{a} = \begin{pmatrix} 6 \\ 2 \end{pmatrix}$ and $\mathbf{b} = \begin{pmatrix} -12 \\ -4 \end{pmatrix}$ are parallel.

Notice that

$$\begin{pmatrix} -12 \\ -4 \end{pmatrix} = -2 \begin{pmatrix} 6 \\ 2 \end{pmatrix}$$

That is $\mathbf{a} = -\dfrac{1}{2}\mathbf{b}$

Since the vector **a** is a scalar multiply of **b**, then **a** and **b** are parallel.

Try this 14

Determine whether or not the given vectors are parallel

(a) $a = \begin{pmatrix} 2 \\ 3 \end{pmatrix}$ and $b = \begin{pmatrix} 4 \\ 5 \end{pmatrix}$,   (b) $a = \begin{pmatrix} -2 \\ 3 \end{pmatrix}$ and $b = \begin{pmatrix} 4 \\ -6 \end{pmatrix}$

Perpendicular Vectors

If two vectors $a = \begin{pmatrix} a_1 \\ a_2 \end{pmatrix}$ and $b = \begin{pmatrix} b_1 \\ b_2 \end{pmatrix}$ are perpendicular then $a_1 b_1 + a_2 b_2 = 0$.

Example

Determine whether or not the vectors $a = \begin{pmatrix} 4 \\ 3 \end{pmatrix}$ and $b = \begin{pmatrix} -3 \\ 4 \end{pmatrix}$ are perpendicular.

We have

$$4(-3) + (3)(4) = -12 + 12 = 0$$

So, the vectors **a** and **b** are perpendicular

Try this 15

Determine whether or not the given vectors are perpendicular

(a) $a = \begin{pmatrix} -2 \\ -1 \end{pmatrix}$ and $b = \begin{pmatrix} 3 \\ -6 \end{pmatrix}$ (b) $a = \begin{pmatrix} -4 \\ 3 \end{pmatrix}$ and $b = \begin{pmatrix} 3 \\ -4 \end{pmatrix}$

Exercise 3.8(a)

1. Which of the following vectors are parallel to $a = \begin{pmatrix} 2 \\ -6 \end{pmatrix}$?

(a) $\begin{pmatrix} -6 \\ 18 \end{pmatrix}$ (b) $\begin{pmatrix} -9 \\ 3 \end{pmatrix}$ (c) $\begin{pmatrix} 10 \\ -30 \end{pmatrix}$ (d) $\begin{pmatrix} 1.5 \\ -4.5 \end{pmatrix}$ (e) $\begin{pmatrix} 8 \\ -12 \end{pmatrix}$

2. Which of the following vectors are perpendicular to $a = \begin{pmatrix} -8 \\ 4 \end{pmatrix}$?

(a) $\begin{pmatrix} 2 \\ 4 \end{pmatrix}$ (b) $\begin{pmatrix} -4 \\ 8 \end{pmatrix}$ (c) $\begin{pmatrix} 3 \\ 6 \end{pmatrix}$ (d) $\begin{pmatrix} -24 \\ 16 \end{pmatrix}$ (e) $\begin{pmatrix} -6 \\ -12 \end{pmatrix}$

Some results in geometry can be verified by using vectors as illustrated by the example below.

Example

The vertices of triangle ABC are A(2,3), B(-4, 1) and C(4, -1). X and Y are the midpoint of AB and AC respectively. Show that vector $\overrightarrow{XY}$ is parallel to $\overrightarrow{BC}$.

First, draw and label the triangle

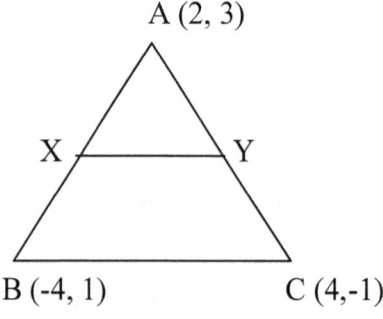

A (2, 3)

X   Y

B (-4, 1)   C (4,-1)

The coordinates of X is $\left(\frac{2+(-4)}{2},\frac{3+1}{2}\right)$ i.e. $(-1,2)$

The coordinates of Y is $\left(\frac{2+4}{2},\frac{3+(-1)}{2}\right)$ i.e. $(3,1)$

$$\overrightarrow{XY} = \overrightarrow{OY} - \overrightarrow{OX}$$

$$= \binom{3}{1} - \binom{-1}{2}$$

$$= \binom{4}{-1}$$

Also $\overrightarrow{BC} = \overrightarrow{OC} - \overrightarrow{OB}$

$$= \binom{4}{-1} - \binom{-4}{1}$$

$$= \binom{8}{-2}$$

You can see that $\overrightarrow{XY} = \frac{1}{2}\overrightarrow{BC}$

Therefore, $\overrightarrow{XY}$ is parallel to $\overrightarrow{BC}$.

This result can also be obtained as follows:

$$\overrightarrow{XY} = \overrightarrow{XA} + \overrightarrow{AY}$$

$$= \frac{1}{2}\overrightarrow{BA} + \frac{1}{2}\overrightarrow{AC}$$

$$= \frac{1}{2}\left(\overrightarrow{BA} + \overrightarrow{AC}\right)$$

$$= \frac{1}{2}\overrightarrow{BC}$$

Hence, $\overrightarrow{XY}$ is parallel to $\overrightarrow{BC}$.

Try this 16

The vertices of an isosceles triangle are A(2,3), B(- 3, - 2) and C(3, - 4). If P is the midpoint of BC, show that AP is perpendicular to BC.

Exercise 3.8(b)

1. A (1, - 2), B (6, 1) and C (3, 6) are the vertices of a triangle. Show

   that the triangle is a right angled triangle.

2. A (- 3, 2), B (1, 4) and C (5, 2) are the vertices of a triangle. P and

   Q are the midpoint of AB and AC respectively. What is the

   relationship between $\overrightarrow{PQ}$ and $\overrightarrow{BC}$.

3. The vertices of a triangle are A (3, - 4), B (- 5, 0) and C (3, 6).

   Show that ABC is isosceles triangle.

4. The vertices of a parallelogram are P (2, 1), Q (2, 5), $R(x, y)$ and

   S (- 2, 2). Find the values of x and y.

5. P (1, 1), $Q(x, y)$, R (1, 3) and S (-1,2) are the vertices of a square.

   Show that PR is perpendicular SQ.

Review exercise 3

1. Find

(a) $\begin{pmatrix} 3 \\ -4 \end{pmatrix} + \begin{pmatrix} -2 \\ 1 \end{pmatrix}$ (b) $\begin{pmatrix} -7 \\ 3 \end{pmatrix} + \begin{pmatrix} 4 \\ -5 \end{pmatrix}$ (c) $\begin{pmatrix} 3 \\ 2 \end{pmatrix} + \begin{pmatrix} -5 \\ -1 \end{pmatrix}$

(d) $\begin{pmatrix} 5 \\ 3 \end{pmatrix} + \begin{pmatrix} -4 \\ -5 \end{pmatrix}$ (e) $\begin{pmatrix} -8 \\ 3 \end{pmatrix} + \begin{pmatrix} 5 \\ -7 \end{pmatrix}$ (f) $\begin{pmatrix} 1 \\ -2 \end{pmatrix} + \begin{pmatrix} -3 \\ 2 \end{pmatrix}$

2. Find

(a) $\begin{pmatrix} 5 \\ 2 \end{pmatrix} - \begin{pmatrix} 6 \\ 3 \end{pmatrix}$ (b) $\begin{pmatrix} -3 \\ 2 \end{pmatrix} - \begin{pmatrix} -4 \\ -2 \end{pmatrix}$ (c) $\begin{pmatrix} 2 \\ 1 \end{pmatrix} - \begin{pmatrix} -2 \\ 4 \end{pmatrix}$

(d) $\begin{pmatrix} 3 \\ -1 \end{pmatrix} - \begin{pmatrix} 5 \\ -2 \end{pmatrix}$ (e) $\begin{pmatrix} 0 \\ -3 \end{pmatrix} - \begin{pmatrix} -5 \\ -3 \end{pmatrix}$ (f) $\begin{pmatrix} -2 \\ -5 \end{pmatrix} - \begin{pmatrix} -3 \\ -7 \end{pmatrix}$

3. Find

(a) $-2 \begin{pmatrix} -1 \\ 3 \end{pmatrix}$ (b) $3 \begin{pmatrix} 2 \\ -4 \end{pmatrix}$ (c) $-4 \begin{pmatrix} 2 \\ -1 \end{pmatrix}$

(d) $\frac{1}{2} \begin{pmatrix} -6 \\ 4 \end{pmatrix}$ (e) $-\frac{3}{2} \begin{pmatrix} -4 \\ 2 \end{pmatrix}$ (f) $\frac{3}{4} \begin{pmatrix} -8 \\ 12 \end{pmatrix}$

4. If $a = \begin{pmatrix} 3 \\ 1 \end{pmatrix}$, $b = \begin{pmatrix} -2 \\ 3 \end{pmatrix}$ and $c = \begin{pmatrix} -4 \\ -2 \end{pmatrix}$, find

(a) $2a + 3b$ (b) $-3a + b$ (c) $2c - b$ (d) $3b - 2a$

(e) $2c - 3a$ (f) $\frac{3}{2}c - a$ (g) $2a + b - 3c$

5. Find the magnitude of the following vectors

(a) $\begin{pmatrix} -4 \\ 3 \end{pmatrix}$ (b) $\begin{pmatrix} 15 \\ -8 \end{pmatrix}$ (c) $\begin{pmatrix} -6 \\ -8 \end{pmatrix}$ (d) $\begin{pmatrix} 5 \\ -12 \end{pmatrix}$

6. Find to the nearest degree the acute angle between the following vectors and the $x - axis$

(a) $\begin{pmatrix} 4 \\ 3 \end{pmatrix}$ (b) $\begin{pmatrix} 5 \\ 2 \end{pmatrix}$ (c) $\begin{pmatrix} 4 \\ -5 \end{pmatrix}$ (d) $\begin{pmatrix} -3 \\ -4 \end{pmatrix}$

7. Express the following vectors as column vectors

(a) (20 units, 050^0)   (b) (15 units, 120^0)   (c) (10 units, 215^0)

8. Find the magnitude- bearing form of the following vectors

(a) $\begin{pmatrix} -5 \\ 12 \end{pmatrix}$ (b) $\begin{pmatrix} 6 \\ -8 \end{pmatrix}$ (c) $\begin{pmatrix} -8 \\ -15 \end{pmatrix}$ (d) $\begin{pmatrix} 4 \\ 3 \end{pmatrix}$

9. Given $\overrightarrow{PQ} = \begin{pmatrix} -3 \\ 4 \end{pmatrix}$ and $\overrightarrow{PR} = \begin{pmatrix} 2 \\ -1 \end{pmatrix}$, find $\overrightarrow{QR}$.

10. Given $\overrightarrow{AB} = \begin{pmatrix} 3 \\ -4 \end{pmatrix}$ and $\overrightarrow{CB} = \begin{pmatrix} -5 \\ 2 \end{pmatrix}$, find $\overrightarrow{AC}$.

11. If A(-5, 2) and B(2, - 3), find $\overrightarrow{AB}$ and $\overrightarrow{BA}$

12. Given P(- 3,2) and $\overrightarrow{PQ} = \begin{pmatrix} 7 \\ -5 \end{pmatrix}$, find the coordinates of Q

13. Given Q(2, - 1) and $\overrightarrow{PQ} = \begin{pmatrix} 5 \\ -2 \end{pmatrix}$, find the coordinates of P.

14. Determine whether or not the given pair of vectors are parallel

(a) $a = \begin{pmatrix} -2 \\ -3 \end{pmatrix}$ and $b = \begin{pmatrix} 5 \\ 4 \end{pmatrix}$ (b) $a = \begin{pmatrix} 2 \\ -3 \end{pmatrix}$ and $b = \begin{pmatrix} -4 \\ 6 \end{pmatrix}$

15. Determine whether or not the given pair of vectors are perpendicular

(a) $a = \begin{pmatrix} -4 \\ 3 \end{pmatrix}$ and $b = \begin{pmatrix} 3 \\ 4 \end{pmatrix}$ (b) $a = \begin{pmatrix} -5 \\ -3 \end{pmatrix}$ and $b = \begin{pmatrix} -2 \\ 4 \end{pmatrix}$

16. A (5, - 1), B (3, 5), C (-3, 3) and D (- 1, - 3) are four points in the plane. Determine the relationship between the following pairs of vectors

(a) $\overrightarrow{AB}$ and $\overrightarrow{DC}$ (b) $\overrightarrow{AC}$ and $\overrightarrow{BD}$

Chapter Test 3

Take this test as you would take a test in class. After you are done, check your work against the answers in the back of the book.

1. Given $a = \begin{pmatrix} -3 \\ 2 \end{pmatrix}$ and $b = \begin{pmatrix} 5 \\ -3 \end{pmatrix}$, find

(a) $3a + 2b$ (b) $2a - 3b$

2. The vertices of a triangle PQR are P(- 4,5), Q(2,- 3) and R(- 4, -5)

(a) Express $\overrightarrow{PQ}, \overrightarrow{QR}$ and $\overrightarrow{PR}$ as column vectors

(b) Show that triangle PQR is isosceles

3. If $a = \begin{pmatrix} 3 \\ 2 \end{pmatrix}$, $b = \begin{pmatrix} 2 \\ 5 \end{pmatrix}$, $c = \begin{pmatrix} 8 \\ 9 \end{pmatrix}$ and $ka + mb = c$, where k and m are real numbers, calculate the values of k and m.

4. $\overrightarrow{AB} = (17\ km, 152°)$ and $\overrightarrow{BC} = (13\ km, 337°)$

(a) Express $\overrightarrow{AB}$ and $\overrightarrow{BC}$ as column vectors

(b) Using your results in (a) find $\overrightarrow{AC}$ in component form

(c) Express $\overrightarrow{AC}$ in magnitude- bearing form

5. ABCD is a parallelogram with vertices $A(x, y)$, $B(3,2)$, $C(8,5)$ and

$D(0,1)$

(a) Find $\overrightarrow{AB}$ and $\overrightarrow{DC}$ and hence find the values of x and y.

(b) Calculate the magnitude of $\overrightarrow{AC}$, correct to three significant figures

4

Construction and Loci

4.1 Construction

Construction is the drawing of plane figures such as lines, angles and triangles accurately, using a pair of compasses and a straight edge. These are the only instruments allowed. The ruler is often used as a straightedge.

Copying Angles

Start with an angle BAC that we will copy.

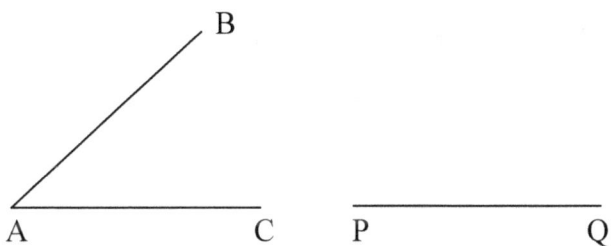

First draw a straight line PQ. This will become one side of the new angle.

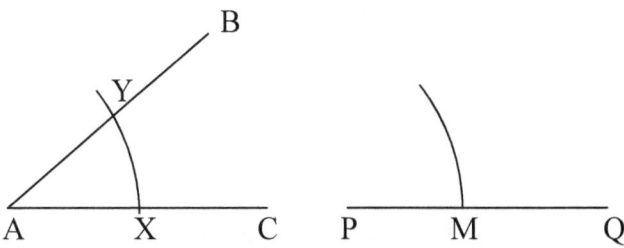

Open up a pair of compasses to any convenient radius. Place the compass point on A and draw an arc to cut AC at X and AB at Y. With the same radius, place the compass point on P and draw an arc to cut PQ at M. Place the compass point on X and adjust its width to point Y. Without changing the compasses width, and with centre at M, draw an arc to cut the first one at L.

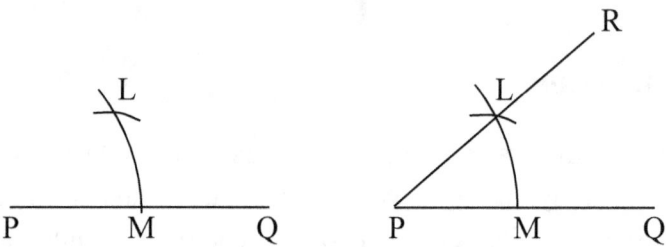

Finally, draw a line PR, from P through L.

Exercise 4.1(a)

1. Draw an angle of any size and then copy it

2. Using a protractor, draw the following angles and then copy them

(a) 50^0 (b) 74^0 (c) 108^0 (d) 136^0 (e) 165^0

Bisecting Angles

Start with angle PQR that we will bisect. Open a pair of compasses to any convenient radius. Place the compass point on Q and draw an arc to intersect QP at X and QR at Y.

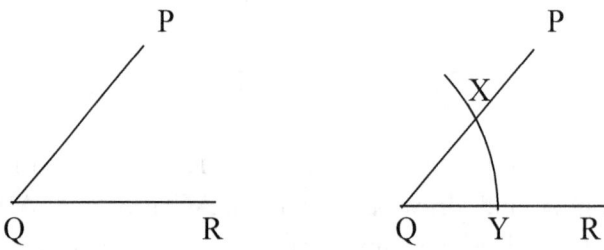

Place the compass point on X and draw an arc between the two arms of the angle. With the same radius, place the compass point on Y and draw an arc to intersect the first arc. Draw a line through Q and the point of intersection of the two arcs. The line bisects angle PQR.

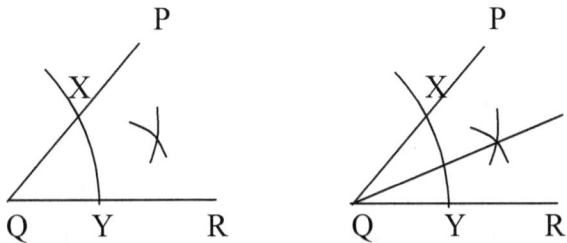

Exercise 4.1(b)

1. Draw an angle of any size and then bisect it, using a pair of compasses

2. Draw the following angles. Bisect the angles using a pair of compasses

(a) 60^0 (b) 76^0 (c) 108^0 (d) 120^0

Perpendicular Bisector of a Line Segment

Start with a line segment PQ.

P   Q

Open the pair of compasses to more than half the distance PQ. Place the compass point on P and draw arcs above and below PQ.

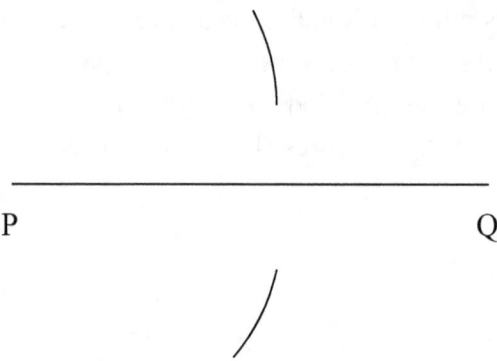

With the same radius, place the compass point on Q and draw arcs above and below PQ so that the arcs intersect the first two arcs (see diagram below).

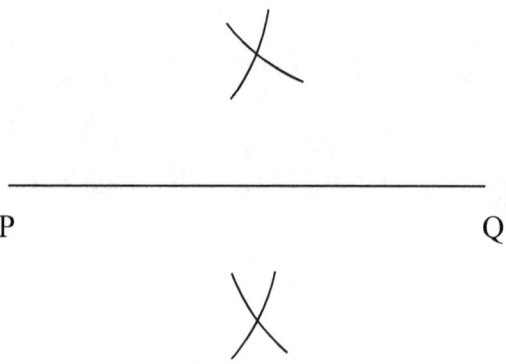

Using a straightedge, draw a line through the two points where the arcs intersect.

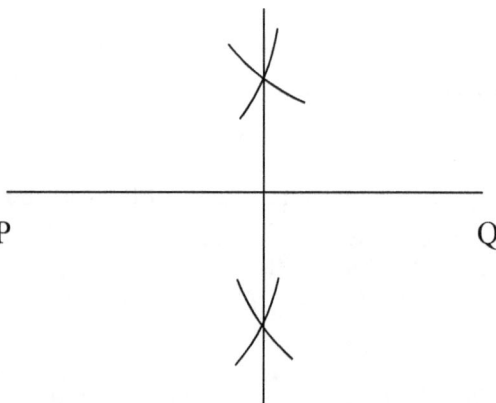

The line drawn is called the perpendicular bisector of PQ. The line meets PQ at right angles and divides it in half.

Exercise 4.1(c)

1. Draw a line of any length and bisect it

2. Draw lines with the following lengths and bisect them

(a) 3 cm   (b) 4.6 cm   (c) 5.2 cm   (d) 6 cm

Constructing Angles

Perpendicular at a point on a line

Begin by drawing a line. Mark a point A on the line

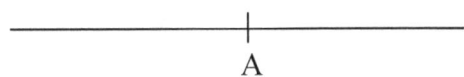

A

Open up a pair of compasses to any convenient radius, place the compass point on A and draw a short arc on the line at each side of the point A, as shown in the diagram below

P   A   Q

Place the compass point on P, and with a radius more than the length PA, draw an arc above PQ. With the same radius place the compass point on Q, and draw an arc to intersect the first arc at R

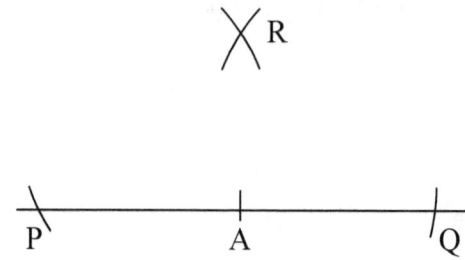

Using a straightedge, draw a line from A through R, (see diagram below).

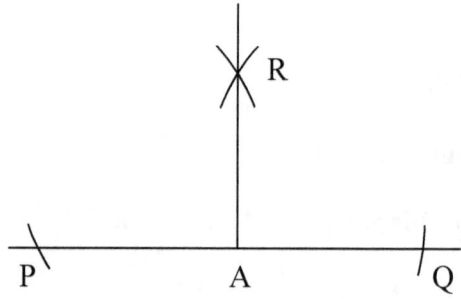

Note: When A is the end of the line, extend the line to enable you draw an arc on each side of A.

Perpendicular at the end point of a line

Begin by drawing the line AB

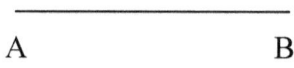

A   B

Open a pair of compasses to any convenient radius. Place the compass point on A, and draw an arc to intersect AB at P. With the same the radius, place the compass point on P, and draw an arc to intersect the first arc at Q. Without changing the radius place the compass point on Q, and draw an arc to intersect the first arc at R (see diagram below).

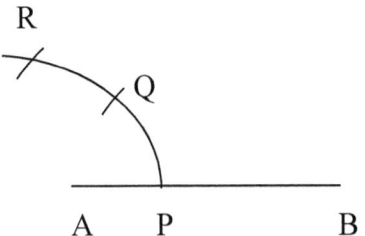

Place the compass point on Q, and draw an arc above the original arc. With the same radius, place the compass point on R, and draw an arc to cross the first arc at S (see diagram below).

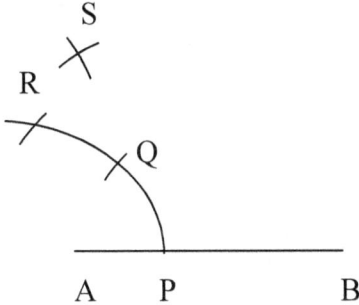

Using a straightedge, draw a line from A through S, as shown in the diagram below.

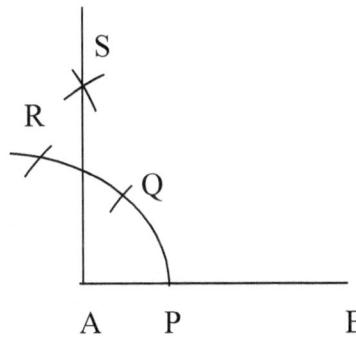

Perpendicular to a line from an external point

Begin with a line and point P which is not on that line

P
•

Place the compass point on P, and with any convenient radius draw an arc to intersect the line at A and B, as shown in the diagram below.

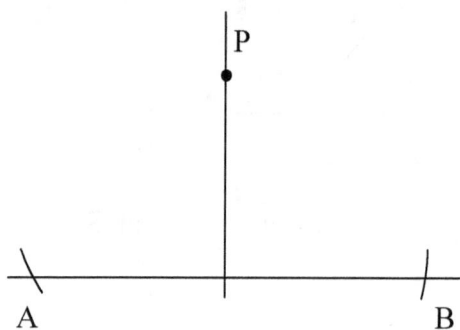

From each point A and B, draw an arc below the line so that the arcs intersect at R.

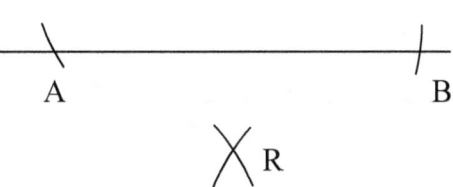

Using a straightedge, draw a line from P through R, as shown in the diagram below.

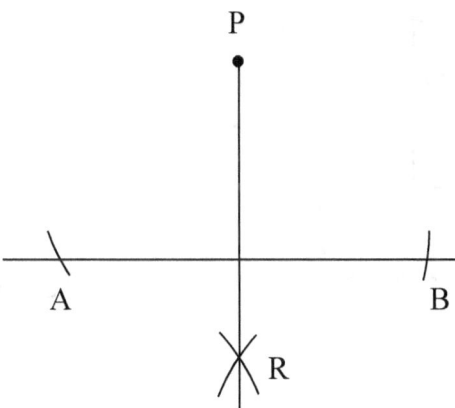

Constructing 45⁰ and 135⁰ angles

Begin by constructing a 90⁰ angle

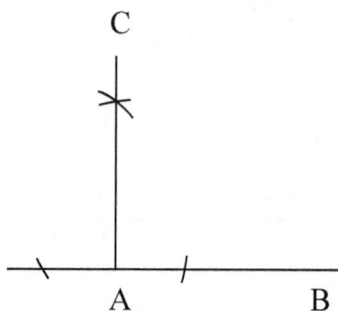

Place the compass point on A, draw an arc to intersect AB at P and AC at Q. From each point P and Q, draw an arc so that the arcs intersect at R, as shown in the diagram below. Draw a line from A to R.

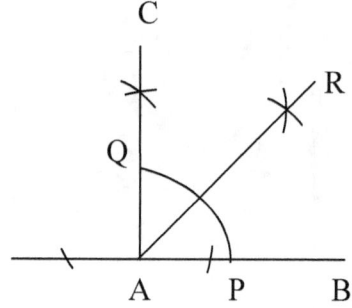

You can construct 135^0, if you draw the arc on the opposite side of AB.

Constructing 60^0 angle

Begin by drawing the line PQ.

P   Q

Draw an arc to intersect PQ at A.

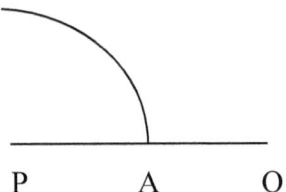

P   A   Q

With the same radius, place the compass point on A, and draw an arc to intersect the first arc at B (see diagram below).

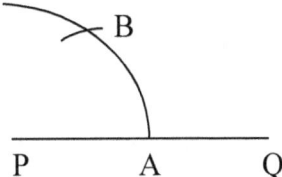

P   A   Q

Draw a line from P through B.

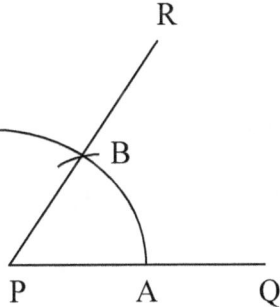

P   A   Q

Constructing 30⁰ angle

Begin by constructing 60⁰.

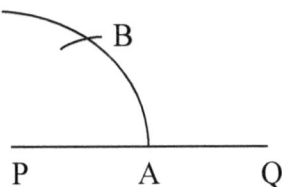

P   A   Q

From each point A and B, draw an arc so that the arcs cross at R, as shown in the diagram below.

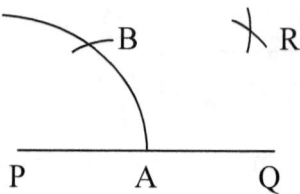

Draw a line from P through R.

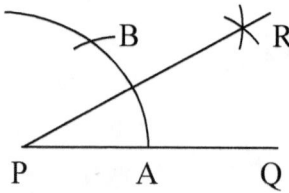

Constructing 75⁰ and 105⁰ angles

Begin by drawing a perpendicular at P, on the line PQ.

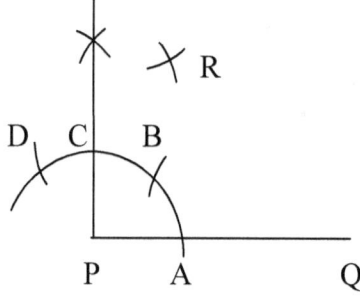

From each point B and C, draw an arc so that the arcs intersect at R.

Draw a line from P through R (see diagram below).

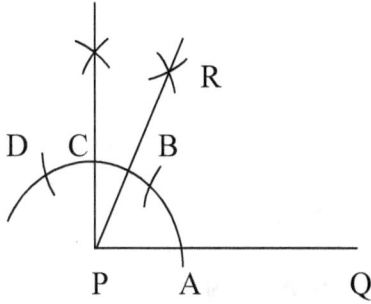

You can construct 105^0, when you draw the arcs from C and D.

Constructing 150^0 angle

Begin by drawing a line. Mark a point P on the line. Place the compass point on P, and draw an arc to intersect the line at A and B, as shown in the diagram below. With the same radius, place the compass point on B, and draw an arc to intersect the first arc at C. Place the compass point on C, and draw an arc to cross the first arc at D.

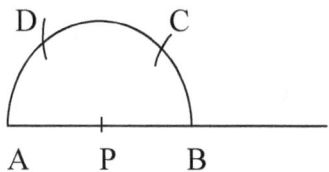

From each point A and D, draw an arc so that the arcs intersect at R. Draw a line from P through R.

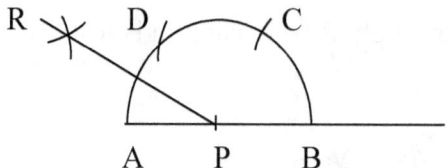

Constructing Parallel Lines

Constructing a parallel line a given distance from a line

Draw a line and mark two points A and B on that line. Construct perpendiculars through these points. Set your compasses radius to the given distance. Using A and B as centres draw arcs to cross the perpendiculars at P and Q. Draw a line through P and Q. (see diagram below)

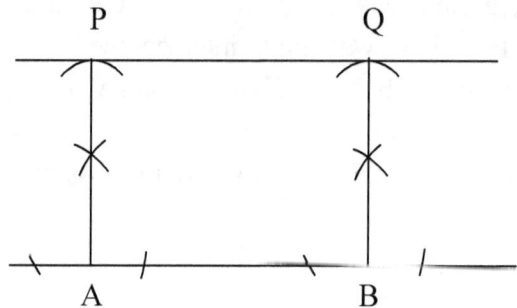

Constructing a parallel line through a point

Start with a line segment AB and a point P not on the line

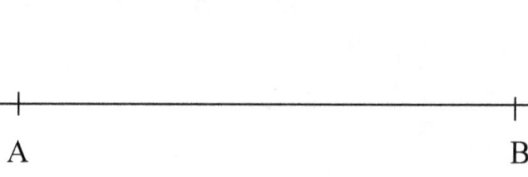

Draw a line through P that intersect AB at Q, as shown in diagram below

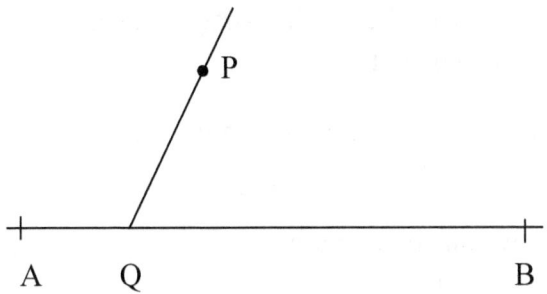

Copy angle PQB at P, as shown in the diagram below

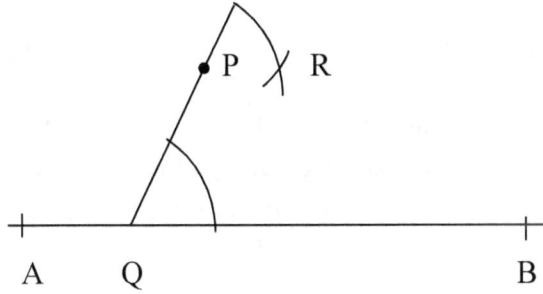

Draw a straight line through points P and R, as shown in the diagram below

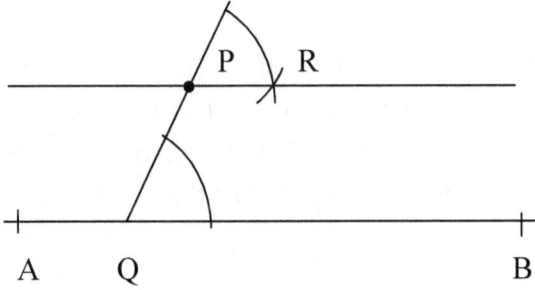

Constructing Triangles

The following steps will be very useful in constructing triangles

1. Make a rough sketch of the finished diagram as it will appear when the construction is completed.

2. With the aid of your sketch and any necessary steps make the construction.

3. Do not erase any construction arcs.

Constructing a triangle given three sides

Example

Construct triangle ABC, AB = 8.5 cm, AC = 7 cm and BC = 9 cm

Begin by constructing the line AB. Draw a line and mark a point A on that line. Place a pair of compasses on a ruler and open up the compasses to 8.5 cm. Place the compass point on A and draw an arc to intersect the line. Label the point of interception B.

Open up the pair of compasses to 7 cm. Place the compass on A and draw an arc above AB. Then open up the pair of compasses to 9 cm. Place the compass point on B and draw another arc to intersect the first arc. Label the point of intersection of the two arcs C. Join A to C and B to C.

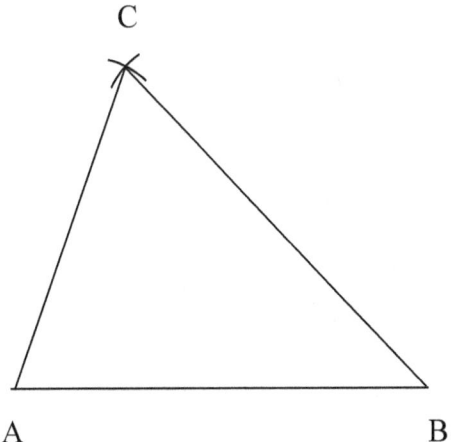

Exercise 4.1(d)

Using only a ruler and a pair of compasses, construct $\triangle ABC$ where:

1. AB = 4.8 cm, BC = 3.6 cm, AC = 6.0 cm

2. AB = 7.5 cm, BC = 4.5 cm, AC = 10 cm

3. AB = 5.6 cm, BC = 4.8 cm, AC = 5.4 cm

Constructing a triangle given two sides and an angle

Examples

Construct triangle ABC where AB = 6 cm, AC = 7 cm and $\angle BAC = 45°$.

Begin by constructing AB = 6 cm. Construct angle 45^0 at A. With A as centre and radius 7 cm, draw an arc to intersect the other arm of the angle 45^0 at C. Join B to C.

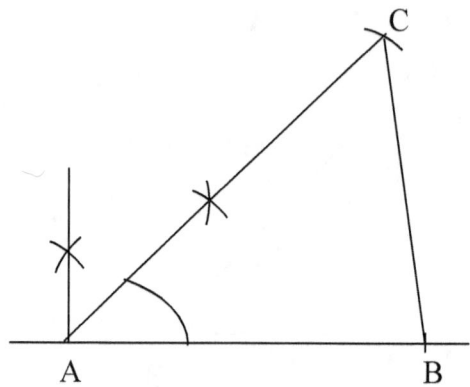

Exercise 4.1(e)

Using only a ruler and a pair of compasses, construct $\triangle ABC$ where:

1. AB = 7 cm, AC = 6 cm, $\angle BAC = 120°$

2. AB = 8 cm, BC = 6.5 cm, $\angle ABC = 75°$

3. AB = 7.5 cm, AC = 6.8 cm, $\angle BAC = 45°$

4. AB = 8 cm, AC = 7cm, $\angle BAC = 60°$

Construct triangle ABC, AB = 6.8 cm, BC = 5 cm and $\angle BAC = 30°$

Begin by constructing AB = 6.8 cm. Construct $30°$ at A. With B as centre and radius 5 cm, draw an arc to intersect the other arm of the angle $30°$ at C. Join B to C.

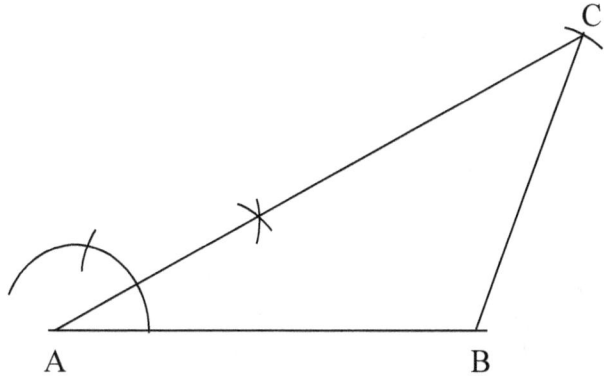

Exercise 4.1(f)

Using only a ruler and a pair of compasses, construct $\triangle ABC$ where:

1. AB = 6.8 cm, BC = 9.7 cm, $\angle BAC = 105°$

2. AB = 7.6 cm, AC = 8 cm, $\angle ABC = 60°$

3. AB = 6.4 cm, AC = 10.5 cm, $\angle ABC = 135°$

Constructing a triangle given one side and two angles

Example

Construct triangle ABC, AB = 7 cm, $\angle BAC = 75°$ and $\angle ABC = 60°$

Begin by constructing AB = 7 cm. At A, construct angle 75^0 and at B construct angle 60^0. Draw the arms of the angles to intersect at C.

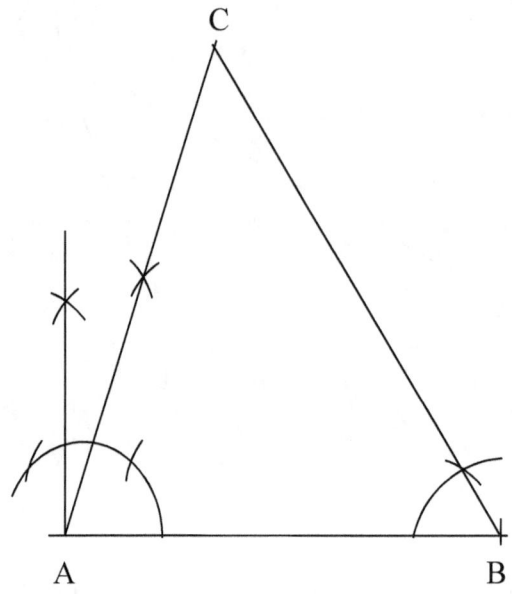

Exercise 4.1(g)

Using only a ruler and a pair of compasses, construct $\triangle ABC$ where:

1. AB = 7 cm, $\angle ABC = 60°$, $\angle BAC = 75°$

2. AB = 7.8 cm, $\angle ABC = 30°$, $\angle BAC = 105°$

3. AB = 6.8 cm, $\angle ABC = 45°$, $\angle BAC = 90°$

Constructing Quadrilaterals

Constructing a square

Example

Construct a square with side 6cm

Begin by constructing the line AB = 6 cm. Construct a perpendicular AP to AB at A. With centre A and radius 6 cm, draw an arc to intersect AP at D. Through D construct DQ parallel to AB. Through

B construct BC parallel to AD to intersect DQ at C. The completed construction is shown below.

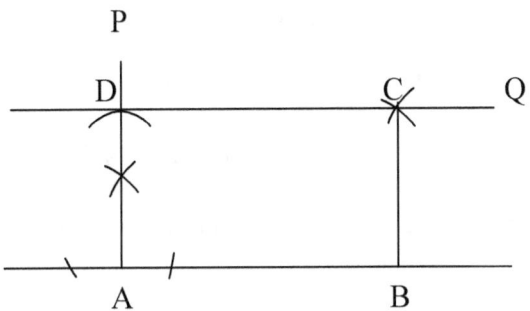

Constructing a parallelogram

Example

Construct a parallelogram ABCD such that AB = 8 cm, BC = 6 cm and $\angle ABC = 120°$.

Begin by constructing the line AB = 8 cm. At B, construct $\angle ABP = 120°$. With centre B and radius 6 cm, draw an arc to intersect BP at C. With centre C and radius 8 cm, draw an arc on the left side of C. With centre A and radius 6 cm, draw an arc to intersect the previous arc at D.

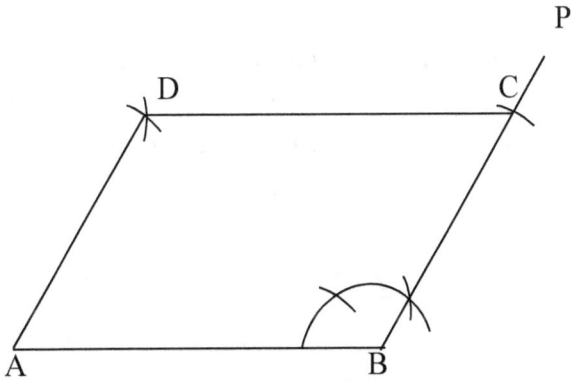

Exercise 4.1(h)

1. Construct a quadrilateral ABCD such that AB = 8 cm, BC = 6 cm, CD = 9 cm, $\angle ABC = 105°$ and $\angle ACD = 45°$

2. Construct a square ABCD with side 5.6 cm

3. Construct a rectangle ABCD such that AB = 4.5 cm and BC = 6 cm.

4. Construct a parallelogram ABCD, such AB = 7.6 cm, AD = 4.8 cm and BD = 10 cm

5. Construct a trapezium ABCD such that AB = 10 cm, AD = 5.4 cm, DC = 7.5 cm, $\angle BAD = 90°$ and AB//DC.

4.2 Loci

A locus is the path of a moving point satisfying a set of given conditions. All points on the locus satisfy these conditions. The plural of locus is loci, pronounced low -sigh.

A locus may be point(s), lines, circles or any combination of these. We will consider four simple loci. Each of the loci can be constructed using a pair of compasses.

1. Locus of a point a given distance from a fixed point

The locus of points which are r unit from a fixed point O, is a circle with centre O and radius r. Recall that all points on the circumference of a circle are equidistant from the centre.

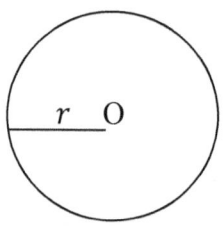

2. Locus of a point equidistant from two given points

The locus of points which are the same distance from two fixed points A and B is the perpendicular bisector of the straight line joining A to B.

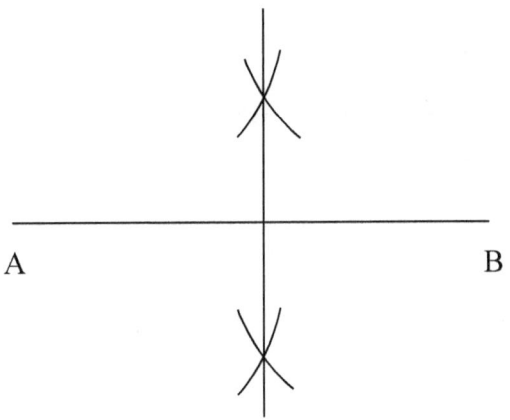

3. Locus of a point equidistant from two fixed lines joined at a vertex

The locus of points which are the same distance from two fixed lines OA and OB is the bisector of the angle formed where the two lines meet, i.e. angle AOB.

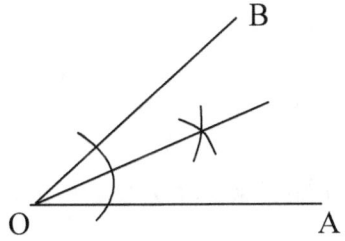

4. Locus of a point a distance r from a line

The locus of all points that are distance r from a given line AB is a pair of parallel lines, one on each side of the line AB, and each a distance r from the line AB.

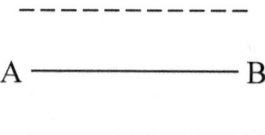

Note: the diagram excludes the end points of the line

Example

Construct triangle ABC with AB = 6 cm, AC = 7 cm and $\angle BAC = 60°$. Construct:

(a) The locus l_1 of points equidistant from AC and BC

(b) The locus l_2 of points equidistant from A and C

(c) The locus l_3 of points 3.5 cm from B.

Begin by constructing triangle ABC.

(a) Construct the angle bisector of angle BCA.

(b) Construct the perpendicular bisector of the line AC

(c) Construct a circle with centre B and radius 3.5 cm

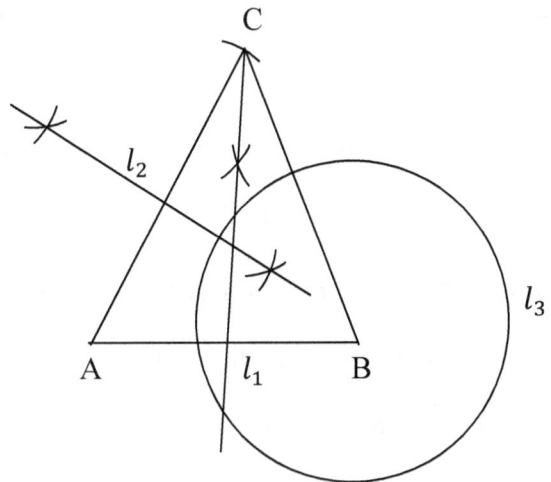

Exercise 4.2

1. Using a ruler and a pair of compasses only

(a) Construct a triangle ABC such that AB = 7.5 cm, AC = 8 cm and ∠BAC = 60°.

(b) Construct the locus l_1 of points equidistant from points A and B to meet AB at P_1.

(c) Construct the locus l_2 of points equidistant from AB and AC

(d) Find the points of intersection, P_2 of l_1 and l_2 and measure

∠P_1P_2A

2. Using a ruler and a pair of compasses only construct

(a) triangle ABC with AB = 7 cm, AC = 8 cm and ∠BAC = 75°

(b) the locus l_1 of points which are equidistant from A and B

(c) the locus l_2 of points 4 cm from B

(d) Find the points of intersection P_1 and P_2 of l_1 and l_2 and measure

$|P_1 P_2|$

3. Using a ruler and a pair of compasses only construct triangle ABC

with AB = 8 cm, $\angle BAC = 45°$ and $\angle ABC = 60°$

(a) Find, by construction a point P inside the triangle which is

equidistant from AB and BC and 5 cm from A

(b) Construct the locus through P of points equidistant from AB to

meet BC at D. Measure $\angle BPD$

4. Using a ruler and a pair of compasses only

(a) Construct a parallelogram PQRS such that PQ = 8 cm, PS = 6

cm and $\angle PQR = 120°$

(b) Locate a point O inside PQRS such that it is equidistant from SR,

SP and PQ.

(c) Construct the circle which touches SR, SP and PQ

(d) Measure the radius of the circle

5. Using a ruler and a pair of compasses only,

(a) Construct a quadrilateral ABCD in which AB = 7 cm, BC = 3.5

cm, AC = 8 cm, BD = 9 cm and $|AD| = |DC|$.

(b) Measure AD.

(c) Find the area of $\triangle ACD$

Scale Drawing

It is not always possible to draw on paper the actual size of real-life objects. A drawing can be created to scale to represent an object which is too large to be drawn on a drawing sheet.

A scale drawing is a drawing that shows a real object with accurate sizes reduced or enlarged by a certain scale. Common uses of scale drawings include maps and plans.

Example

The diagram below shows a triangular plot of land. B is 65 metres from A, and 70 metres from C. C is 80 metres east of A.

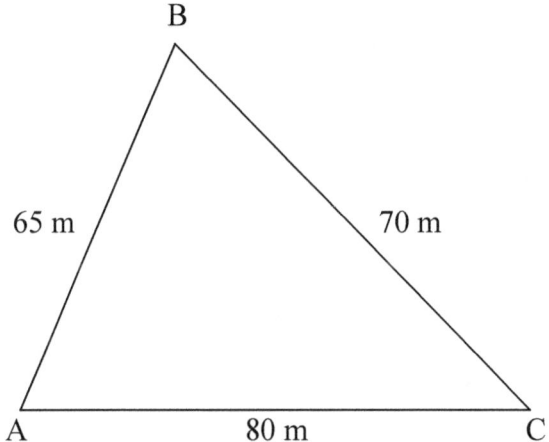

(a) Using a scale of 1 cm for each 10 m, construct a scale diagram of the triangular plot. Use a straight edge and a pair of compasses only.

(b) Mark a point P on the diagram which is equidistant from A, B and C.

First, calculate the length of the scale diagram.

$$\frac{Length\ of\ drawing}{Actual\ length} = \frac{1}{10}$$

$$\frac{Length\ of\ AB}{65} = \frac{1}{10}$$

$$Length\ of\ AB = \frac{1}{10} \times 65$$

$$= 6.5\ \text{cm}$$

Similar proportion gives $AC = 8$ cm and $BC = 7$ cm.

The diagram is shown below. The point P is the point of intersection of the perpendicular bisector of AB and the perpendicular bisector of AC.

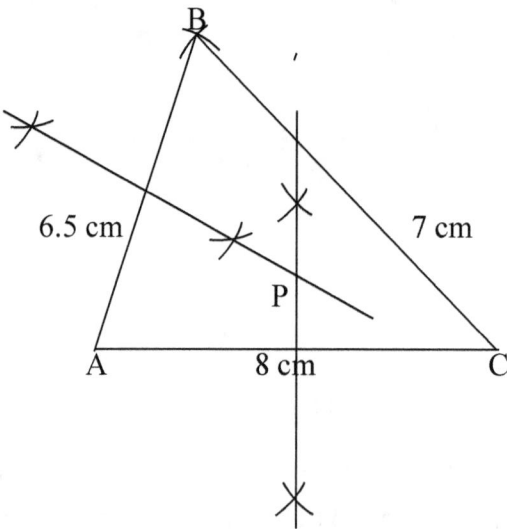

Exercise 4.3

1. A plan view of a farm is shown below

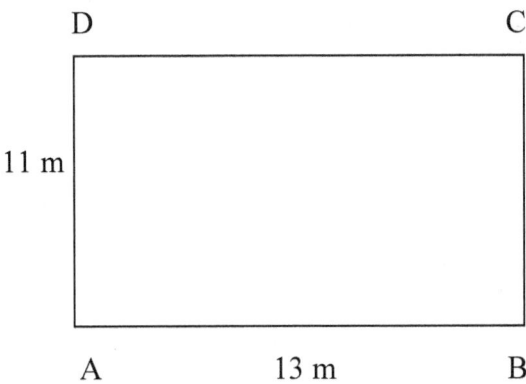

(a) Using a scale of 1 cm to 2 m, construct a scale drawing of the farm

(b) A farmer wants to construct a fishpond on the farm. He wants to construct it not more than 10 m from B, and closer to A than B. Shade the part on the farm where the fishpond could be constructed

2. P and Q are two towns. Q is 70 kilometres directly east of P. A telecommunication company is building a mast and because of the terrain would locate the mast 50 kilometres from P and 60 kilometres from Q.

(a) Draw a scale diagram to show the above information, using a scale of 1 cm to 10 km.

(b) By construction locate the two points that could be used as a possible site for the mast.

3. A rectangular field measures 25 m by 45 m as shown below

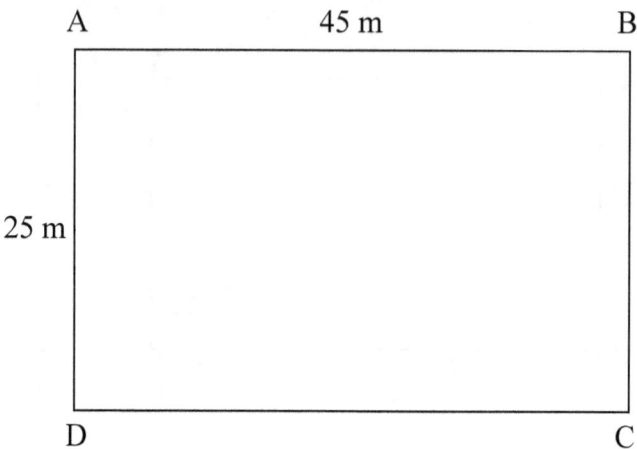

(a) Make a scale drawing of the field, using 1 cm to represent 5 m.

(b) A boy walk across the field in such a way that his path is the perpendicular bisector of BD. Show by construction, the path that he takes

(c) Calculate the distance the boy walk

4. A triangular plot is shown on the diagram below

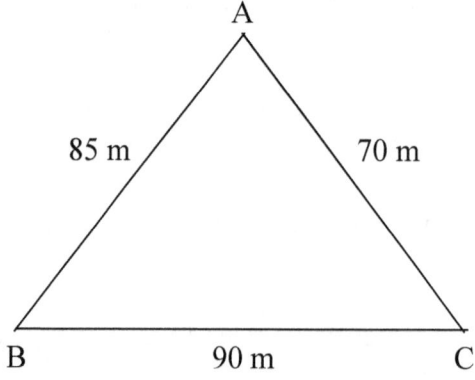

(a) Make a scale drawing of the plot, using a scale of 1 cm to 10 m

(b) Shade the region inside the triangular plot which is closer to AB than CA, and greater than 40 m from A

5. The diagram shows the position of three villages P, Q and R.

P is 12 kilometres north of Q and 20 kilometres west of R.

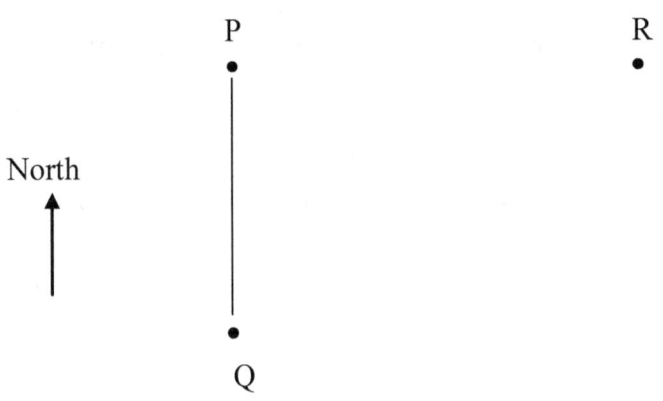

(a) Make a scale drawing of the diagram, using a scale of 1 cm to 2 kilometres

(b) Show by construction the position of a school, S, which is equidistant from P and Q, and the bearing of the school from Q is 045^0

(c) Find the actual distance of the school from R

6. A and B are two points on a coastline. S is a ship at sea. B is 36 kilometres due East of A, and 20 kilometres from S. The bearing of S from A is 060^0.

(a) Construct a scale drawing of the diagram, using a scale of 1 cm to 4 kilometres

(b) The ship sails to the coastline on a course that is an equal distance from SA and SB. Using a ruler and compasses only, construct the course of the ship

Review exercise 4

In Exercises 1 – 12, use only ruler and compasses.

1. Construct a triangle ABC such that AB = 7 cm, BC = 8 cm, $\angle ABC = 60°$. Construct the bisector of AB and AC. Mark the point of intersection of the bisectors O. With a radius OA and centre O, construct a circle. Measure OA.

2. Construct a triangle ABC such that AB = 8 cm, AC = 6.5 cm and $\angle BAC = 105°$. Construct the bisectors of $\angle ABC$ and $\angle BAC$. Mark the point of intersection of the bisectors O. Construct the perpendicular from O to AB to meet AB at P. With radius OP and centre O, construct a circle. Measure OP.

3. Construct triangle ABC with AB = 8 cm, $\angle BAC = 75°$ and $\angle ABC = 30°$. Construct the bisector of AC and mark the point its meet AC, P. Measure $\angle ABP$

4. Construct triangle ABC with $AB = AC = 7.5$ cm and $\angle BAC = 60°$ Construct the perpendicular from C to meet AB at P. Construct the bisector of $\angle ABC$ and mark its point of intersection with CP as Q. With PQ as radius and centre Q, draw a circle. Measure PQ.

5. Construct triangle ABC with $AB = 6$ cm, $BC = 8.5$ cm and $\angle BAC = 90°$. Construct the bisector of $\angle BAC$ to meet BC at P. Measure $\angle APB$. Describe $\triangle APB$. Construct a line through P parallel to AB to meet AC at Q. Measure AQ.

6. Construct a quadrilateral ABCD such that $\angle BAC = 45°$, $\angle ABC = 60°$, $AB = DC = 8$ cm and AB//DC. Measure $\angle ADC$

7. Construct a quadrilateral ABCD with $AB = 7$ cm, $AC = 7.5$ cm, $\angle ABC = 75°$, $\angle ADC = 60°$ and $AD = CD$. Measure AD.

8. Construct a rhombus ABCD with $AB = 6.3$ cm and $AC = 10$ cm. Construct the bisector of $\angle ADC$ to meet AC at P. Measure $\angle DPC$. Construct the perpendicular from P to DC to meet DC at Q. With PQ as radius and centre P construct a circle. Measure PQ.

9(a) Construct an equilateral triangle ABC with sides of lengths 7 centimetres.

(b) Construct the locus l_1, of points equidistant from A and C. Mark the point P of the point of intersection of l_1 and AC.

(c) Construct a straight line l_2 through P which is parallel to AB.

(d) Construct the locus l_3 of points 3.5 cm from P

(e) Mark the two points Q and R which are equidistant from AB and BC, and 3.5 cm from P. Join QR and describe the figure APQR.

10(a) Construct triangle PQR with PQ = 7 cm, QR = 9 cm and $\angle QPR = 75°$

(b) Construct the locus l_1 of points equidistant from P and Q.

(c) Construct the locus l_2 of points equidistant from PQ and QR.

(d) Construct the locus l_3 of points 5 cm from P.

(e) Shade the region inside the triangle which is closer to Q than P, further from the line PQ than the line QR, and within 5 cm of P

11. P, Q and R are three villages. R is 16 kilometres directly east of Q. P is 12 kilometres from Q and 25 kilometres from R.

(a) Construct a scale diagram to show the above information, using a scale of 1 cm to 2 km.

(b) The villages want to place a fire station equidistant from all three villages. By construction, locate the point that satisfies the requirement.

12. A and B are two radar stations on a coastline. A is 12 kilometres directly west of B. A ship S, 15 kilometres from A and on a bearing of $030°$ from A, sails to the cost.

(a) Draw a scale diagram to show the above information, using a scale of 1 cm to 2 km.

(b) Station A could pick signals up to 7 kilometres away and Station B could pick signals up to 8 kilometres away. If the ship sends a distress call, shade the possible area in which the

distress call could be picked by any of the radar station.

Chapter Test 4

Take this test as you would take a test in class. After you are done, check your work against the answers in the back of the book

Use only ruler and pair of compasses

1(a) Construct triangle ABC with sides AB = 6 cm, AC = BC = 7 cm.

(b) Construct a straight line through C to bisect AB at P. Measure

$\angle APC$

(c) Bisect AC

(d) Locate the point Q which is equidistant from A and C, and also

from AC and BC.

(e) Construct the perpendicular from Q to BC to meet BC at R.

(f) With QR as radius and centre Q construct a circle. Measure QR.

2. The diagram below shows a plot of land in the shape of a

quadrilateral ABCD with sides AB = 100 km, AD = 60 km,

DC = 80 km, BC = 120 km, and $\angle BAD = 150°$

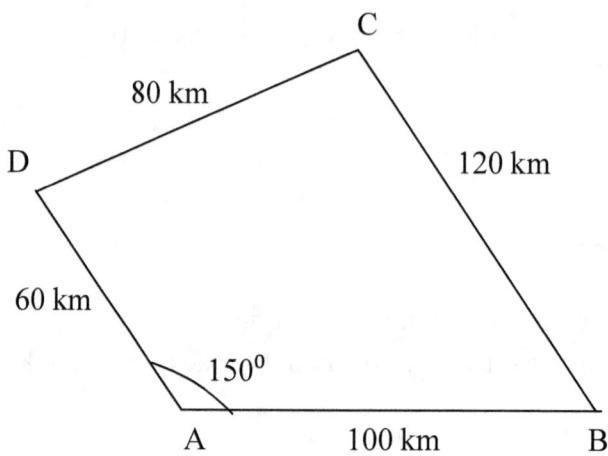

Draw the quadrilateral accurately, using a scale of 1 cm to 10 km.

A man wants to plant a lawn which is closer to C than D, further from the side DA than the side AB, and within 70 kilometres of D. Shade the part on the plot where the lawn could be planted.

5

Probability

Most events in life have uncertain outcomes. For example, we may not know in advance whether it will rain tomorrow or not.

Probability of an event is a measure of how likely it is that the event will occur. Probability can assume a value between 0 and 1, inclusive. Some events will never occur, and others will always occur. The probability of an event that is impossible is 0 and the probability of an event that is certain is 1. Any other event will have a probability lying within the range 0 to 1. If an event is very unlikely to occur, the probability would be close to 0. If an event is very likely to occur, the probability would be close to 1.

The probability of an event can be expressed as a fraction, percent or decimal.

Basic Probability Terms

Experiments

An experiment is an action where the result is uncertain. For example, tossing a coin, rolling a die or drawing a card from a pack of cards are all experiments.

When you draw a ball from a bag containing 10 balls, each individual draw is called a trial. The result of a single trial is an outcome. For example, if a single die is roll, all the possible outcomes are 1, 2, 3, 4, 5 or 6.

Sample Space

All the possible outcome of an experiment is called a sample space. For example, the sample space when a die is rolled is {1, 2, 3, 4, 5, 6}.

Events

An event is one or more outcome of an experiment. For example, getting a tail when you toss a coin or rolling an even number.

The possible outcome of an experiment is usually known but which outcome will actually occur is unknown. An event that actually happens is called a favourable (or successful) outcome.

Definition of Probability

If a fair coin is tossed, a head (H) is as likely to come up as a tail (T). That is, the outcomes are equally likely to occur.

The probability that a head comes up in a single toss of a coin, denoted by P(H) is $\frac{1}{2}$. Similarly, $P(T) = \frac{1}{2}$.

In general, if there are n equally likely events, the probability of each one happening is $\frac{1}{n}$.

Example

A fair die is rolled. Find the probability of rolling a 3.

Recall that possible outcomes are 1, 2, 3, 4, 5 or 6, so the number of possible outcome is 6. The number of favourable outcome is 1.

The probability of rolling a 3 is $\frac{1}{6}$

Try this 1

A spinner has the numbers 1 – 10 find the probability of spinning a 10.

Consider a box containing 9 balls of which 5 are black and 4 are white. If a ball is picked at random, each of the 9 balls is equally likely to be picked. The probability of picking a black ball is $\frac{5}{9}$ as the event of picking a black ball takes place in 5 out of 9 possible outcomes.

If an event E is equally likely then

$$P(E) = \frac{the\ number\ of\ favourably\ outcome}{the\ number\ of\ possible\ outcomes}$$

Example

If a fair die is rolled, what is the probability of rolling an even number?

The possible outcome is 1, 2, 3, 4, 5 or 6, so the number of possible outcome is 6.

The favourable outcome is 2, 4 or 6, so the number favourable outcome is 3

So, the probability of rolling an even number is $\frac{3}{6}$, i.e. $\frac{1}{2}$.

Try this 2

If a card is drawn from a pack of 52 cards, what is the probability of drawing a red card?

Try this 3

(a) A number is chosen at random from set P = {3, 5, 7, 9}. Find the probability that it is:

 (i) even and (ii) odd

(b) A ball is picked at random from a box containing 8 black balls, 7 red balls and 5 white balls. Find the probability of picking

(i) a black ball   (ii) a red ball

(iii) a white ball   (iv) Add all the probabilities

You may have noticed the following properties of probability:

1. If an event E is certain to occur, then $P(E) = 1$

2. If an event E cannot occur, then $P(E) = 0$

3. The probability that an event E will occur lies between 0 and 1.

4. The sum of the probabilities of all possible events equal 1

Complement of an Event

The complement of an event is the event not occurring. The complement of an event A is denoted by A', and read not A.

Since an event and its complement together make all the possible outcomes,

$P(A) + P((A') = 1$, and $P(A') = 1 - P(A)$

Sometimes it would be easier to find the probability that an event will occur by calculating the probability that the event will not occur and then subtract the result from 1.

Example

A die is rolled. The probability of rolling a 5 is $\frac{1}{6}$, what is the probability of not rolling 'a 5'

If A is an event of rolling 'a 5', then

$$P(A') = 1 - \frac{1}{6}$$

$$= \frac{5}{6}$$

Try this 4

A box contains 12 balls which are identical but differ in colour. 4 of the balls are blue and the rest are white. If a ball is drawn at random, what is the probability that it will:

(a) be a blue ball?   (b) not be a blue ball?

Exercise 5.1

1. A die is rolled. What is the probability of rolling a number less than 4?

2. A die is rolled. What is the probability of rolling either a 3 or a 4?

3. A ball is drawn at random from a box containing 6 black balls and 4 white balls. What is the probability of drawing a white ball?

4. A boy chooses a number at random from 1 to 20 inclusive. What is the probability that the number is divisible by 2?

5. A player picks a ball at random from a bag containing 50 tennis balls, 35 of which are white and the rest red. What is the probability that he picks a red ball?

6. In a class of 60 students, 50 of them read chemistry and 40 read physics. What is the probability that a student chosen at random in the class reads both chemistry and physics?

7. A bag contains a number of red and blue balls. The probability of drawing a blue ball is $\frac{7}{12}$. What is the probably of drawing a red ball.

8. A bag contains red and white balls. The probability of picking a red ball is $\frac{1}{4}$. If the number of balls in the bag is 20, find the number of white balls

9. A card is drawn from a pack of 52 cards. What is the probability of drawing a picture card?

10. A card is drawn from a pack of 52 cards. What is the probability of drawing a red card?

11. 500 tickets are sold for a raffle. If the probability that the tickets bought by a man will win the first prize is 0.06, how many tickets did he buy?

12. A man bought 12 tickets for a raffle. If the probability that he will win the first prize is 0.02, how many tickets were sold?

5.2 Relative Frequency

If an event is not equally likely to occur, the probability that the event may occur can be estimated from experiments or historical data.

For instance if Ofori and Yeboah played 20 games in a year, and Yeboah won 12 of the games, the probability of Yeboah winning the next game can be estimated by using the formula

$$\frac{number\ of\ successful\ trials}{total\ number\ of\ trials}$$

This ratio is called the relative frequency.

In this case the probability that Yeboah wins the next game is $\frac{12}{20}$ i.e. $\frac{3}{5}$.

The relative frequency of an event is an estimate of the probability. When an experiment is repeated a large number of times the relative frequency provides a good estimate of the probability

Example

The data shows the colour of cars passing a school gate one morning

Colour	Red	Black	Blue	Gray	Others
Frequency	15	8	12	9	16

Estimate the probability that, at a random time, a car passing will be red.

The total number of cars $= 15 + 8 + 12 + 9 + 16 = 60$

The relative frequency of red cars is $\frac{15}{60} = \frac{1}{4}$

So the probability of a red car passing the school gate is $\frac{1}{4}$ or 0.25.

Try this 5

The sides of a six- sided die is rolled 100 times. The table shows the result

Number roll	1	2	3	4	5	6
Frequency	26	19	17	15	14	9

(a) What is the probability of rolling 1?

(b) What is the probability of rolling an old number?

(c) What is the probability of rolling a factor of 6?

Exercise 5.2

1. The ages of 20 students who attended a birthday party are shown in the table below

Ages(in years)	10	11	12	13	14	15	16	17
Number of students	1	2	3	4	5	3	1	1

If a student is picked at random, what is the probability that he will be

(a) 13 years old

(b) at least 14 years old

(c) at most 12 years old

2. The table shows the number of defective articles in a quantity of articles produced in a factory

Number of article produced	100	120	140	160	180	200
Number of defective articles	15	16	18	20	25	26

If an article is picked at random what is the probability that it will be defective?

3. The following table shows the grades obtained by students in a mathematics examination

Grades	A	B	C	D	E	F
Number of students	15	20	25	20	24	16

If a student is picked at random what is the probability that he obtained

(a) grade D

(b) grade C or better

(c) a grade lower than D

4. The distribution of first year students in a school is shown in the table below

Class	Number of students
General Art	45
Business	50
Science	30
Technical	35
Vocational	40

If a student is chosen at random, what is the probability that

(a) he is in the business class

(b) he is in the Science or Technical class

5. The table shows the number of students absent from school for 40 days

Number of students	0	1	2	3	4	5
Number of days	12	8	5	6	5	4

If a day is chosen at random what is the probability that

(a) at least 3 students will be absent

(b) all the students will be present

6. A school entered candidates for the WASCE examination from 1992 – 1996. The number who qualified to write the university entrance examinations are shown in the table below.

Year	1992	1993	1994	1995	1996
Number qualified	96	145	155	130	124

If the school entered 200 candidates every year, what is the probability that a candidate chosen at random will qualify to write the entrance examination.

5.3 Probability of Combine Events

Recall that an event can include one or more possible outcomes. Combined probability involves two or more events. Before we discuss the laws of probability, we state the following definitions:

Mutually Exclusive Events

Two events are mutually exclusive if the events cannot happen at the same time.

Independent Events

Two events are independent if the probability of one event is not affected by the occurrence of the previous event.

Dependent Events

Two events are dependent if the probability of one event is affected by the occurrence of the previous event. For example, if you pick two red balls one after the other without replacement, that is without putting the first ball back, from a bag containing 4 red balls and 6 black balls then the probability of picking a red ball in the second draw is $\frac{3}{9}$ i.e. $\frac{1}{3}$

Notice that the number of balls reduces by 1 and the number of red balls also reduces by 1.

Laws of Probability

Addition Law of Probability

If two events are mutually exclusive then the probability of any one event occurring is the sum of the individual probabilities. For instance, if A and B are two mutually exclusive events, then

$$P(A \ or \ B) = P(A) + P(B)$$

The symbol ∪ can be used for 'or'.

Example

A 6 -sided die is rolled. What is the probability of rolling a 3 or a 5?

The two events are mutually exclusive, since rolling a 3 and a rolling a 5 cannot occur at the same time.

$$P(3 \ or \ 5) = P(3) + P(5)$$

$$= \frac{1}{6} + \frac{1}{6}$$

$$= \frac{1}{3}$$

Try this 6

A day of the week is chosen at random. What is the probability of choosing a Tuesday or Friday?

Exercise 5.3(a)

1. A 6 sided die is rolled, what is the probability of rolling a 2 or a 5?

2. A 6 sided die is rolled, what is the probability of rolling a 3 or an even number?

3. A 6 sided die is rolled, what is the probability of rolling a 2 or an odd number?

4. A number from 1 to 10 is chosen at random. What is the probability of choosing a 5 or an even number?

5. A roulette wheel has the numbers 1 to 20. What is the probability that the number which comes up will be either 6 or 15?

6. A box contains 6 red, 5 green and 4 blue balls. If a ball is taken from the box at random, what is the probability that it is a red or a blue ball?

7. A box contains 7 white, 8 yellow and 5 black balls. If a ball is taken from the box at random, what is the probability that it is a white or a yellow ball?

8. A card is drawn at random from a pack of 52 cards; find the probability of drawing a card that is either a spade or a diamond?

9. A card is drawn at random from a pack of 52 cards; find the probability of drawing a card that is either a king or a jack.

10. A card is drawn at random from a pack of 52 cards. What is the probability that it is either a King or a Queen?

11. Of the households in a town, 25 % have a car and 30 % have a bicycle. Find the probability that a household has either a car or a bicycle but not both.

12. In a group of students it is known that 25 % have breakfast and 40 % have lunch. Find the probability that a person chosen at random from this group has either breakfast or lunch but not both.

Multiplication Law of Probability

If two events are independent, then the probability of both events occurring is the product of the individual probabilities. For instance if A and B are two independent events then

$P(A \ and \ B) = P(A) \times P(B)$

The symbol ∩ can be used for 'and'.

Example

Two fair dice are rolled. What is the probability that one die gives a 2 and the other die a 3?

If we let A represents the event 'a die gives a 2' and B represents the event 'a die gives a 3', then

$P(A \ and \ B) = P(A) \times P(B)$

$$= \frac{1}{6} \times \frac{1}{6}$$

$$= \frac{1}{36}$$

Try this 7

Two fair dice are rolled. Find the probability that both are sixes.

Exercise 5.3(b)

1. In a throw of two fair dices, what is the probability that one die gives a 5 and the other die an even number?

2. Two cards are drawn from a well- shuffled pack of 52 cards. If the first card is replaced before the second card is drawn, find the probability that both are hearts.

3. A bag contains 12 red and 8 black marbles. Two marbles are drawn. If the first marble is replaced before the second marble is drawn, find the probability that both are red.

4 A bag contains 6 blue and 4 green counters. Two counters are drawn. If the first counter is replaced before the second counter is drawn, find the probability that the first is blue and the second is green.

5. The probability that a man will pass a certain examination is $\frac{3}{5}$ and the probability that his wife will pass the same examination is $\frac{2}{3}$. What is the probability that they will all pass the examination?

6. Two people take a driving test. One has probability of $\frac{2}{3}$ of passing, and the other has probability of $\frac{3}{4}$ of passing, find the probability that they all pass.

7. A marksman hits a target with probability $\frac{3}{5}$. What is the probability of him getting one hit followed by one miss?

8. A box contains 40 textbooks including 5 mathematics textbooks. If a book is picked at random from the box at a time, replacing the first book before the second book is picked; find the probability that both are mathematics textbook.

9. Two archers each independently fire an arrow at a target. The probability that the first archer hits the target is $\frac{3}{5}$ and the probability that the second archer hits the target is $\frac{4}{9}$. Find the probability that one of them hit the target.

10. A boy loves car toys. The probability his father buys him a car toy for his birthday is $\frac{3}{5}$., and the probability that his mother buys him a car toy is $\frac{7}{12}$. What is the probability that he receives one car toy for his birthday?

Non- exclusive Events

Two events are non- mutually exclusive if they have one or more outcomes in common.

If A and B are two non- mutually exclusive events then the probability of one or both of the two events happening is the sum of the probabilities of each event less the probability of both events happening together,

That is $P(A \cup B) = P(A) + P(B) - P(A \cap B)$, if $A \cap B \neq \emptyset$

Example

A single card is drawn at random from a pack of 52 cards. What is the probability of drawing a heart or a picture card?

A pack of 52 cards has 13 hearts and 12 picture cards. Three of the cards are both hearts and a picture card. If A represents the event of drawing a heart and B represents the event of drawing a picture card, then

$$P(A \cup B) = P(A) + P(B) - P(A \cap B)$$

$$= \frac{13}{52} + \frac{12}{52} - \frac{3}{52}$$

$$= \frac{11}{26}$$

Try this 8

A single card is drawn at random from a pack of 52 cards. What is the probability of drawing either an Ace or a Spade?

Exercise 5.3(c)

1. Two fair dice are rolled. What is the probability of rolling at least one odd number?

2. Two fair dice are rolled. What is the probability that either of them will be a 2?

3. Two fair dice are rolled. What is the probability of rolling a 2 or an even number?

4. A card is drawn from a pack of 52 cards. What is the probability that it is either a picture card or red card?

5. A card is drawn from a pack of 52 cards. What is the probability that it is a Queen or a black card?

6. A single card is chosen at random from a pack of 52 playing cards. What is the probability of choosing a King or a Club?

7. The probability that two marksmen will hit a target are $\frac{2}{5}$ and $\frac{5}{8}$ respectively. What is the probability that at least one of them will hit the target?

8. In a mathematics class of 40 students, 23 are boys and 17 are girls. On a unit test, 9 boys and 10 girls made an A grade. If a student is chosen at random from the class, what is the probability of choosing a girl or an A student?

9. The probability that a day is wet is $\frac{1}{5}$ and that it is windy is $\frac{1}{3}$. The probability that it is both wet and windy is $\frac{1}{6}$. Find the probability that it is either wet or windy.

10. The probability that fufu is on the menu for lunch is $\frac{2}{3}$, and the probability of beans $\frac{1}{4}$. The probability of both being on the menu is $\frac{1}{5}$. What is the probability that either fufu or beans will be on the menu.

11. The probability that a man will pass a certain examination is $\frac{3}{5}$ and the probability that his wife will pass the same examination is $\frac{5}{8}$. What is the probability that either the man or his wife pass the examination?

12. The probability that a father will attend his son's graduation is $\frac{3}{5}$ and the probability that the mother will attend is $\frac{7}{12}$. What is the probability that either the father or mother will attend the son's graduation?

Conditional Probability

The probability that an event occurs given that some other event has occurred is called a conditional probability. The conditional probability of event A, given that event B had occurred is denoted by the symbol $P(A/B)$.

If A and B are two dependents events then the probability that both events occur is

$$P(A \cap B) = P(A) \cdot P(B/A)$$

Example

Two cards are selected at random from a pack of 52 cards. If the second card is selected without replacing the first card, what is the probability that two aces are selected?

The probability of selecting an ace on the first draw is $\frac{4}{52}$. The probability of selecting the second ace is $\frac{3}{51}$, since there are now 3 aces out of the remaining 51 cards.

The probability of selecting two aces is $\frac{4}{52} \cdot \frac{3}{51} = \frac{1}{221}$

Try this 9

Two balls are drawn from a box containing 4 blue and 5 red balls. If the first ball is not replaced, what is the probability that the first ball drawn is blue and second is red?

Exercise 5.3(d)

1. Two cards are drawn from a well- shuffled pack of 52 cards. If the first card is not replaced before the second card is drawn, find the probability that both are hearts.

2. Two cards are drawn from a pack of 52 cards. If the first card is not replaced before the second card is drawn, find the probability that the first is a black card and the second is a red card.

3. A bag contains 12 red and 8 black marbles. Two marbles are drawn. If the first marble is not replaced before the second marble is drawn, find the probability that both are red.

4. A bag contains 6 blue and 4 green counters. Two counters are drawn. If the first counter is not replaced before the second counter is drawn find the probability that the first is blue and the second is green.

5. Two cards are drawn from a pack of 52 cards. If the first card is not replaced before the second card is drawn, find the probability that the second card is a Queen.

6. A box contains 15 textbooks including 5 mathematics textbooks. Two books are picked from the box at random. If the first book is not replaced before the second book is picked, find the probability that both books are mathematics textbook.

7. Two letters are chosen from the word SUCCESS. What is the probability that both C's are chosen?

8. Two people are chosen from 10 women and 8 men. What is the probability that they are both men?

9. A class has 12 girls and 14 boys. Two students are selected at random find the probability that the first is a girl and the second is a boy

10. Two different words are chosen from the word VERDICT. Find the probability that one is a vowel.

5. 4 Using Probability Diagrams

Possibility Diagrams

One way of illustrating probabilities for combined events is to list all the possible outcomes.

Example

Two six sided fair dice, a red die and a blue die, are rolled. Find the probability that both numbers are prime numbers.

The table shows all possible ordered pairs in the sample space. Notice that there are 36 outcomes in all.

		1	2	3	4	5	6
				Blue die			
	1	(1,1)	(1,2)	(1,3)	(1,4)	(1,5)	(1,6)
	2	(2,1)	**(2,2)**	**(2,3)**	(2,4)	**(2,5)**	(2,6)
Red die 3	3	(3,1)	**(3,2)**	**(3,3)**	(3,4)	**(3,5)**	(3,6)
	4	(4,1)	(4,2)	(4,3)	(4,4)	(4,5)	(4,6)
	5	(5,1)	**(5,2)**	**(5,3)**	(5,4)	**(5,5)**	(5,6)
	6	(6,1)	(6,2)	(6,3)	(6,4)	(6,5)	(6,6)

The outcomes for which both numbers are prime numbers are shown in bold type.

9 of the 36 outcomes have both numbers prime, so the probability is
$$\frac{9}{36} = \frac{1}{4}$$

Try this 10

Two fair dice are rolled. Find the probability that

(a) the sum is 7   (b) the sum is even   (c) the sum is at most 9

Tree Diagrams

You can use tree diagrams to help solve probability problems involving combined events. Tree diagrams allow us to see all the possible outcomes of an event.

Example

A box contains 8 black balls and 7 red balls. If two balls are drawn without replacement, find the probability that one ball of each colour is drawn.

The tree diagram lists all the possible outcomes.

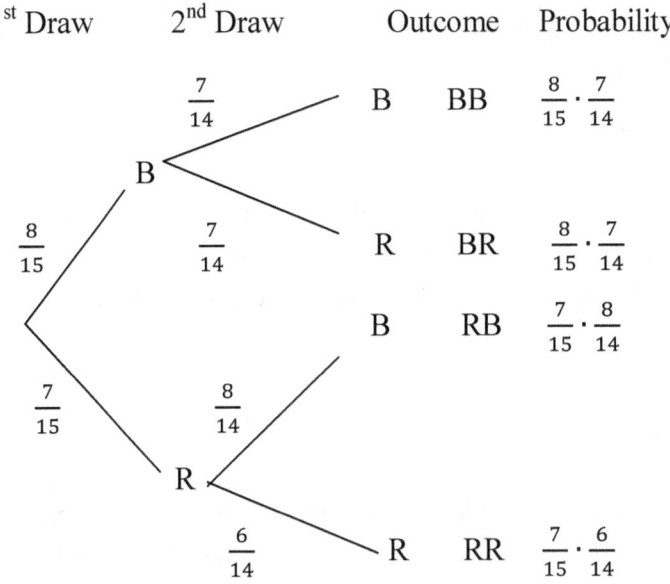

The first draw has two outcomes represented by the two initial branches. The outcome is written at the end of the branch, and the probability of each event is written by the side of the branch. Each of the two initial branches has two other branches.

The probability of the second draw depends on the result of the first draw. The branch from B to B is the conditional probability of drawing a black ball on the second draw after a black ball on the first draw. The probability of drawing a black ball on the second draw is $\frac{7}{14}$, since there are now 7 black balls and a total of 14 balls in the box. The remaining entries are obtained in a similar way.

Notice that the sum of the probabilities for any set of branches is always 1. The probability of any outcome is found by multiplying the probability along the branches.

The event 'drawing one of each colour' is the same as the event 'either first ball drawn is black and the second ball is red or the first ball is red and the second ball is black'. The two events are mutually exclusive so the probability that one ball of each colour is drawn is

$$\frac{8}{15}\cdot\frac{7}{14}+\frac{7}{15}\cdot\frac{8}{14}=\frac{8}{15}$$

Try this 11

A bag contains 3 red marbles and 5 blue marbles. If two marbles are drawn from the bag one after the other without replacement what is the probability that both marbles are of the same colour.

Exercise 5.4

1. Two fair dice are rolled. One of the dice has 6 faces numbered from 1 to 6. The other die has 4 faces numbered from 1 to 4. What is the probability that the total score is

(a) 7

(b) less than 5

2. Two fair coins are tossed simultaneously. What is the probability of obtaining

(a) two tails?   (b) at least one head?

3. Two fair dice are rolled. What is the probability that

(a) at least one six is rolled ?

(b) the total score is 9 ?

(c) the total score is less than 7 ?

4. P = {1, 2, 3, 4} and Q = {5, 6, 7} are two sets of numbers. If two numbers are chosen at random one from each set, find the probability that

(a) both are even

(b) one is odd and the other is even

5. A box contains 9 red and 6 white balls. If two balls are drawn without replacement, what is the probability that both balls are red?

6. A box contains 7 white and 3 black balls. If two balls are drawn at random one after the other without replacement, what is the probability that both balls are of the same colour?

7. A bag contains 9 red and 6 black balls. Two balls are drawn without replacement, what is the probability that they are different colours?

8. A box contains 5 white, 4 blue and 3 green balls. Two balls are drawn at random without replacement; find the probability that one white and one blue ball are drawn.

9. The probability that a man will pass a driving test is $\frac{2}{3}$, and the probability that his daughter will pass the test is $\frac{3}{5}$. Find the probability that if they both take the test one of them will pass.

10. A machine makes two parts which fit together to make a tool. The probability that the first part will be made correctly is 0.9. The probability that the second part will be made correctly is 0.95. Find the probability that both parts will be made correctly.

Review exercise 5

1. A letter is chosen at random from the letters in the word

PROBABILITY, what is the probability that the letter is a vowel?

2. If a letter is picked at random from the alphabet, what is the probability that the letter is not a vowel?

3. If a number between 1 and 20, inclusive is picked at random, what is the probability that the number is a multiple of 2?

4. If a number between 1 and 20, inclusive is picked at random, what is the probability that the number is a perfect square?

5. A card is taken at random from a pack of 52 cards, what is the probability that the card is a picture card?

6. A six sided die is thrown. The table shows the result for 100 throws.

Number thrown	1	2	3	4	5	6
Frequency	18	27	15	17	16	7

What is the probability of throwing a number greater than 3?

7. The table shows the colour of cars in a garage.

Colour	Red	Black	Yellow	Grey	Blue	Others
Frequency	70	15	3	45	52	15

If one is selected at random, what is the probability that its colour is either black or grey?

8. A bag contains balls numbered 1 to 36. If a ball is picked, what is the probability that the ball drawn is numbered either 10 or 30?

9. A 10- sided die has its faces numbered 1 to 10. If it is rolled, what is the probability that it lands on a face numbered 3 or 7?

10. A card is drawn from a pack of 52 cards. What is the probability that it is either an Ace or a King?

11. Two fair dice are rolled. What is the probability that either the first is a 2 or the second is a 5?

12. A card is drawn from a pack of 52 cards. What is the probability that it is either a Spade or a Queen?

13. Two fair dice, A and B, each with faces numbered 1 to 6 are rolled together. What is the probability that the number on die B is a multiple of the number on die A?

14. A first card is drawn at random from a pack of 52 cards and is then put back into the pack. A second card is drawn, what is the probability that the first card is a club and the second is a heart?

15. A marksman hits a target with probability $\frac{4}{5}$. Assuming independence for successive firings, find the probability of getting a miss followed by a hit.

16. One box contains 4 white balls and 2 red balls, and another contains 3 white balls and 5 red balls. If one ball is drawn from each bag, find the probability that one is white and one is red.

17. A card is chosen at random from a pack of 52 cards. Without replacing it, a second card is chosen. What is the probability that the first card chosen is a heart and the second card chosen is a heart?

18. A school survey found that 7 out of 30 students walk to school. If two students are selected at random, what is the probability that all two walk to school?

19. A box contains 4 white balls and 6 black balls. A ball is drawn at random from the box and is then put back into the box. A second ball is then drawn, draw a tree diagram and use this to find the probability that the first ball is white and the second is black.

20. A box contains 8 white balls and 7 yellow balls. If two balls are drawn without replacing the first ball, draw a tree diagram and use it to find the probability of drawing one of each colour.

Chapter Test 5

Take this test as you would take a test in class. After you are done, check your work against the answers in the back of the book.

1. A whole number is picked at random from numbers 1 to 50, inclusive. What is the probability that it is more than 27?

2. A 6 sided die is rolled. What is the probability that a number less than 5 is rolled?

3. The sides of a six sided spinner are numbered from 1 to 6. The table shows the results for 100 spins.

Number of spinner	1	2	3	4	5	6
Frequency	26	19	17	15	16	7

What is the probability of getting a 2 or 5?

4. A bag contains 4 blue, 1 red, 3 black and 2 green pens. What is the probability of taking a black or blue pen?

5. A single card is drawn at random from a pack of 52 cards, what is the probability of drawing either a heart or a picture card?

6. What is the probability of getting just one 6 when a fair die is thrown twice?

7. A box contains 10 cards of which 2 are red and 8 are white. A card is picked at random, its colour is noted and then it is replaced. If two cards are picked at random, what is the probability that the first card is red and the second card white?

8. A bag contains 3 red balls and 2 blue balls. A ball is taken out and not replaced. A second ball is then taken out, what is the probability that the second ball taken is red?

9. Two fair dice are rolled. One of the dice has 6 faces numbered from 1 to 6. The other die has 4 faces numbered from 1 to 4. If the two dice are rolled, what is the probability that the sum of the numbers is more than 5?

10. A bag contains 5 white balls and 3 black balls. A ball is taken out and not replaced. A second is then taken out, draw a tree diagram and use it to find the probability that one of each colour is picked.

11. A box contains 4 red and 2 blue balls. A ball is drawn at random and then replaced. A second ball is then drawn at random. Using a tree diagram find the probability of drawing two balls of the same colour.

12. Bag A contains 10 marbles of which 2 are red and 8 are black. Bag B contains 12 marbles of which 4 are red and 8 are black. A ball is drawn at random from each bag. Draw a tree diagram and use it to find the probability that at least one red ball is drawn.

6

Statistics

Information (called data) such as the number of people living in a country, the number of births and deaths and the number of children in school is collected by the state. A careful analysis of such data enables a government to plan the economic and social development of a country. The science of statistics is concerned with the collection, presentation, analysis and interpretation of data.

6.1 Frequency Distribution

Raw Data

A group of students who attended a camp where asked to write their names and ages on arrival. The ages in years of the first twenty arrivals is shown below:

14, 15, 16, 14, 17, 15, 16, 13, 14, 14, 12, 13, 15, 12, 15, 13, 15, 16, 17, 18

The data has not been written down in any particular order and it is hard to follow and extract useful information from it. The above data is called a raw data.

Frequency

The ages of the twenty students listed above can be arranged in ascending (or descending) order of magnitude, called an array.

12, 12, 13, 13, 13, 14, 14, 14, 14, 15, 15, 15, 15, 15, 16, 16, 16, 17, 17, 18

Notice that the set of data contains many of the same data values. The number of times a particular data value appears is called the frequency. For instance, the frequencies for data values 12 and 16 are 2 and 3 respectively.

Frequency Distribution

The most common way to organize raw data is to construct a table that list all data values in order of magnitude with their corresponding frequencies, as illustrated in Table 6.1.

Tables such as Table 6.1 are called frequency distribution tables.

Ages(in years)	Frequency
12	2
13	3
14	4
15	5
16	3
17	2
18	1

Table 6.1

The frequency distribution table provides a quick summary of the data. Certain trends become more evident. A look at the table clearly shows that 15 is the most frequently occurring age group.

It will be cumbersome to list ages say, from 1 to 100. In this situation we will find it more convenient to distribute the data into groups. These groups are called classes. The groups are arranged so that no value can appear in two groups. The frequency table shown   in Table 6.2 is known as a grouped frequency table or a grouped data.

Ages(in ages)	Frequency
1 – 10	3
11 – 20	4
21 – 30	6
31 – 40	5
41 – 50	2

Table 6.2

Group frequency tables are used to summarize large collection of data so that useful information can be extracted and used.

Class Intervals and Class Limits

A symbol defining a class such as 1 – 10 in the above grouped data is called a class interval. The end numbers, 1 and 10, are called class limits. The smaller number 1 is called the lower class limit and the larger number 10 is called the upper class limit.

Class Boundaries

We could consider the class interval 1 – 10 to include ages from 0.5 to 10.5 years, 0.5 and 10.5 are called the class boundaries or true class limits. The smaller number 0.5 is called the lower class boundary and the larger number 10.5 is called the upper class boundary.

The Size or Width of a Class Interval

The difference between the lower and upper class boundaries is called the class size or class width.

The Class Midpoint or Class Mark

The class midpoint is the midpoint of the class interval and is obtained by adding the lower and upper class limits together and dividing by two. The class midpoint represents all the values in the class.

Forming Frequency Distribution Tables

The steps in constructing frequency tables are illustrated by the following examples below.

Ungrouped Frequency Tables

Example

The ages of 20 students in years are shown below:

14	18	16	18	17
15	19	14	16	17
16	17	17	16	16
17	18	15	17	15

Table 6.3

Form a frequency table for the data

The table is made up of three columns. The headings of the columns are Ages, Tally and Frequency. In the first column we list the ages of the students in ascending order of magnitude. Going through each column of Table 6.3, we place a tally mark, /, by a data value each time it occurs, as shown in the second column of Table 6.4. The total number of tally marks is entered in the third column.

Table 6.4, shows the tally marks for the eight items in the first and second columns of Table 6.3.

Ages(in years)	Tally	Frequency
14	/	
15	/	
16	/	
17	//	
18	//	
19	/	

Table 6.4

The complete tally marks, and the frequency of each data value are shown in Table 6.5.

Ages(in years)	Tally	Frequency
14	//	2
15	///	3
16	////-	5
17	////-/	6
18	///	3
19	/	1

Table 6.5

Notice that each fifth tally mark is made across the first four; this makes it easier to count.

Try this 1

The following are marks of students in a test:

26	28	28	26	27	28
27	29	25	30	26	25
28	26	29	27	27	27
25	27	30	25	29	26
27	26	27	28	26	27

Form a frequency table for the data

Group Frequency Tables

Example

The marks obtained by 30 students in an examination are:

2	45	27	28	13	16
24	25	34	5	59	17
46	7	36	23	48	29
32	39	27	38	16	37
43	25	26	12	36	54

Table 6.6

Form a grouped frequency table, with class intervals 1 – 10, 11 – 20, etc.

We write down the rest of the class intervals. Notice that the last class interval contains the largest number in the data. Tally the values as described in the preceding example.

Table 6.7 shows the completed distribution table.

Marks	Tally	Frequency
1 – 10	///	3
11 – 20	////-	5
21 – 30	////-////	9
31 – 40	////-//	7
41 – 50	////	4
51 – 60	//	2

Table 6.7

Try this 2

The following are the ages (in years) of 30 people in a church:

12	55	37	38	23	26
34	35	44	15	68	27
56	17	46	33	58	38
42	48	37	48	26	47
53	35	36	22	46	64

Form a grouped frequency table, with the class intervals 10 – 19, 20 – 29, etc.

Exercise 6.1

1. The ages (in years) of 36 boys in a group are:

9	8	12	8	9	10
8	13	15	11	12	11
10	14	7	12	10	13
12	12	11	13	12	9
12	11	9	9	14	12
11	10	10	11	10	11

Form a frequency table for the data

2. The marks obtained by 30 students in a test are:

6	8	8	6	7	8
7	9	5	10	6	5
8	6	9	7	7	7
5	7	10	5	9	6
7	6	7	8	6	7

Form a frequency table for the data

3. The heights (in cm.) of 30 members of a club are:

152	158	164	154	169	160
161	179	169	172	178	174
162	153	172	156	162	163
167	168	170	168	162	168
157	163	159	165	153	164

Form a group frequency table, with the group 151 – 155, 156 – 160, etc.

4. The ages (in years) of 40 members of staff of a company are:

32	38	31	45	48	28	47	34
37	23	43	40	33	52	24	25
42	44	39	40	53	37	42	42
33	34	46	24	44	58	48	38
27	29	32	36	32	32	38	27

Form a group frequency table, with the class intervals 21 – 25, 26 – 30, 31 – 35, etc.

5. The marks obtained by 50 students in an examination are:

45	44	64	25	91	71	52	66	52	16
58	44	78	81	64	88	8	85	35	24
33	63	5	75	22	43	95	34	51	41
45	32	56	43	72	34	65	34	23	38
21	55	56	12	52	53	15	62	42	18

Form a group frequency table, with the class intervals 0 – 9, 10 – 19, etc.

6.2 Graphical Representation of Data

Recall that, in Section 6.1, we learned how to organize and summarize data using a frequency distribution table. Graphs can be used as visual representation of data from frequency distribution tables. There are many graphical methods, but in this section we will consider three graphical methods.

Histogram

A histogram consists of a series of rectangular bars joined together and with bases on a horizontal axis. The area of a bar represents the frequency. In the case of bars of equal width, it is customary to take the height of the bar as numerically equal to the frequency.

Examples

The table shows the marks obtained by 20 students in a test marked out of 5.

Marks	0	1	2	3	4	5
Frequency	1	2	4	6	5	2

Draw a histogram to illustrate this information.

Draw two perpendicular axes. Label the vertical axis frequency and the horizontal axis marks. The horizontal axis is labelled according to the information given in the data.

Choose a suitable scale for the vertical axis. Mark intervals of equal length on the horizontal axis. For each interval, construct a bar which height equals to the corresponding frequency. Write the label of each bar at the centre of the bars.

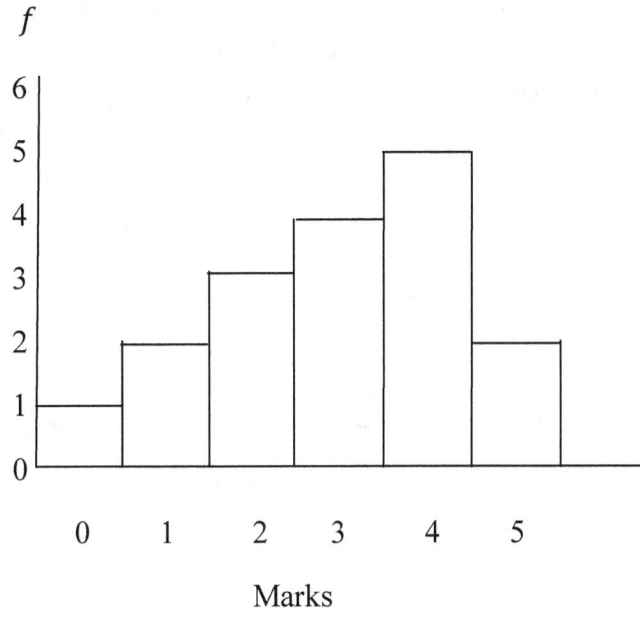

Figure 6.1

Try this 3

The ages (in years) of children in each of 25 families is shown in the table below

Ages (in years)	1	2	3	4	5	6
Frequency	3	5	7	6	3	1

Draw a histogram for the distribution

The ages (in years) of 30 people at a camp are shown in the table below

Ages (in years)	6 - 10	11 - 15	16 - 20	21 - 25	26 - 30	31 - 35
Frequency	2	4	6	8	7	3

Draw a histogram to illustrate this information

All the class intervals are the same, so the base of the histogram will all be of equal width. The histogram is constructed as in the example above. However, there are two ways you can label the bars. You can use the class midpoint or the class boundaries. The class midpoints are marked at the centre of each bar. The class boundaries are marked exactly at the boundaries of the bars as shown in Figure 6.2.

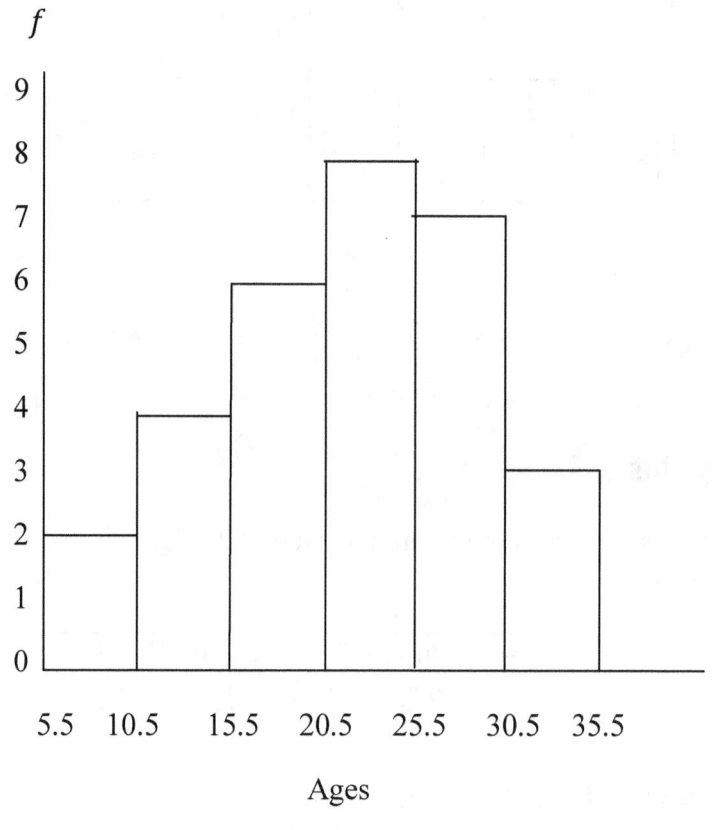

Figure 6.2

Try this 4

The table shows the distribution of marks scored by 20 students in a test. Draw a histogram for the distribution

Marks	20 – 24	25 - 29	30 - 34	35 - 39	40 - 49
Frequency	3	4	7	5	1

Bar Charts

A bar chart consists of vertical rectangular bars drawn with equal width. A bar chart differs from a histogram because the bars are not joined. The gabs between the bars may be equal to the width of the bars.

Exercise 6 2(a)

1. The ages of 20 students in a school football team are shown in the table below

Ages (in years)	14	15	16	17	18
Frequency	3	4	6	5	2

Illustrate this data on a histogram

2. The table below gives the frequency distribution of marks scored by a class in a test

Marks	0	1	2	3	4	5
Frequency	2	3	5	9	7	4

Illustrate this data on a histogram

3. The marks obtained by 40 students in an examination is shown in the table below

Marks	21 - 24	25 - 28	29 - 32	33 - 36	37 - 40	41 - 44	45 - 48	49 - 52
Frequency	4	5	6	7	8	5	3	2

Illustrate this data on a histogram

4. The data below shows the distribution of the heights (in cm.) of workers in a factory.

Height(cm)	164 - 166	167 - 169	170 - 172	173 - 175	176 - 178	179 - 181	182 - 184
Frequency	3	4	5	9	3	4	2

Illustrate this data on a histogram

5. The ages, in years, of people who attended an art exhibition are shown in the table below

Ages	11-15	16-20	21-25	26-30	31-35	36-40	41-45
Frequency	5	7	8	12	9	5	4

Illustrate this data on a histogram

Pie Chart

A pie chart is a circle divided into sectors. The size of each sector is proportional to the frequency of each category. Since a circle contains 360^0, you can obtain the angles of the sectors from the equation

$Angle = Relative\ frequency \times 360°$

Example

The first year students of a school are distributed in the following programs

Art   Science   Business   Vocational   Technical

36   34   40   38   32

Draw a pie chart to illustrate this information

Begin by adding the numbers of categories.

Total number of students $= 36 + 34 + 40 + 38 + 32 = 180$

Next, calculate the size of each sector angle as shown below.

Art: $\frac{36}{180} \times 360° = 72°$

Science: $\frac{34}{180} \times 360° = 68°$

Business: $\frac{40}{180} \times 360° = 80°$

Vocational: $\frac{38}{180} \times 360° = 76°$

Technical: $\frac{32}{180} \times 360° = 64°$

Note: it is worth checking your result by adding your answers.

$72° + 68° + 80° + 76° + 64° = 360°$

Figure 6.3 shows the pie chart constructed from the angles calculated above.

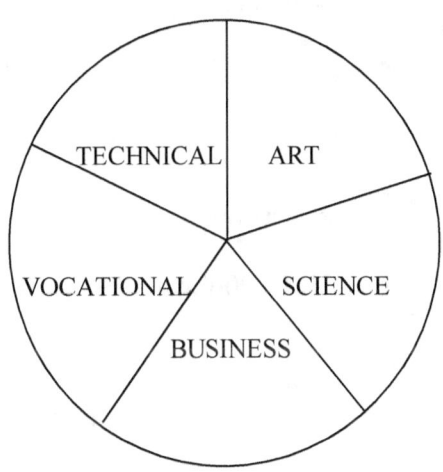

Figure 6.3

Try this 4

The table below shows the number of students who took a test in a subject

Mathematics	Physics	Chemistry	Biology	Economics	History
45	39	28	14	36	18

Draw a pie chart to illustrate the data

Exercise 6.2(b)

1. Draw a pie chart to show the expenditure (in Ghana cedis) of a man in a month

Food	Transport	Clothing	Entertainment	Miscellaneous
474	395	316	237	158

2. The books in a school library are classified as follows:

Fiction	Science	Mathematics	Social Studies	Biography	Reference books
30%	25%	10%	5%	10%	20%

Draw a pie chart to show this distribution

3. A private school receives GH¢15,000 from the following sources

	GH¢
Tuition	3,600
Grants	2,400
Donations	x
Income from school farms	1,600
Interest on endowment found	2,000
Contributions from old students	4,000

(a) Find x

(b) Draw a pie chart to illustrate the following information

4. The pie chart below shows the distribution of marks scored by a student in a test

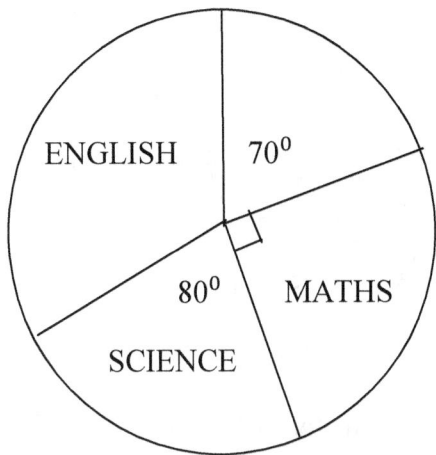

If his total score is 72, find the mark he scored in English

5. The table shows the distribution of students admitted in various programs in a school.

Programs	Arts	Business	Science	Technical	Vocational	Agriculture
No. of students	219	189	189	183	141	154

Construct a pie chart to illustrate the distribution

6. In an election, the number of votes won by five contestants A, B, C, D and E in a village are as follows. Draw a pie chart to illustrate the information.

Contestants	A	B	C	D	E
Number of votes	140	110	190	520	240

7. The pie chart shows how a school allocates its funds to its departments.

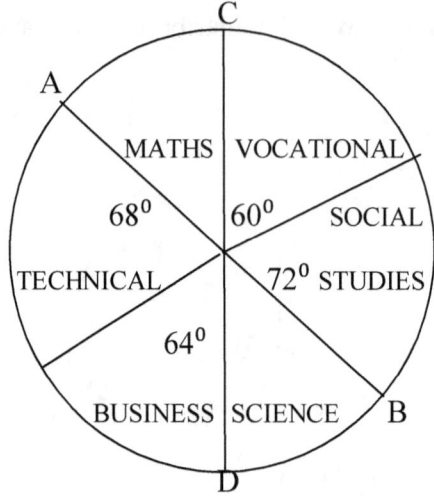

AB and CD are two diameters of the circle. If the school allocates GH¢1,350 quarterly to the Mathematics department

(a) find the total amount allocated to the departments.

(b) what percentage of the fund was allocated to Social Studies and Science?

6.3 Cumulative Frequency Distribution

Marks	Frequency
1 – 5	1
6 – 10	5
11 – 15	9
16 – 20	16
21 – 25	14
26 – 30	3
31 – 35	2

Table 6.8

Table 6.8 shows the marks obtained by 50 students in a test. It may be useful to show the total number of students who obtained marks below or above a certain value. For example, the total number of students who obtained marks less than 21 is $1 + 5 + 9 + 16$ i.e. 31, The total frequency is called the cumulative frequency.

A table displaying cumulative frequencies is called cumulative frequency distribution or cumulative frequency table. Table 6.9 shows the cumulative frequency distribution of the distribution given in Table 6.8.

Marks less than	Cumulative frequency
5.5	1
10.5	6
15.5	15
20.5	31
25.5	45
30.5	48
35.5	50

Table 6.9

The cumulative frequency distribution shown in Table 6.9 is referred to as a less than cumulative frequency distribution.

Note the following:

1. The values in the first column are the upper class boundaries

2. Each entry in the cumulative frequency column is obtained by adding successive frequencies.

3. The last value in the cumulative frequency column is equal to the total frequency

Try this 5

The table shows the height (in cm.) of 40 people in a club

Heights	110 - 114	115 - 119	120 - 124	125 - 129	130 - 134	135 - 139	140 - 144
Frequency	2	4	6	10	8	6	4

Construct a cumulative frequency table for the distribution

Cumulative Frequency Curve (Ogive)

A graph obtained by plotting the cumulative frequency against the upper class boundary and joining all the points by a smooth curve is called a cumulative frequency curve or ogive.

A cumulative frequency curve can be used to determine how many items lie above or below a certain value in a distribution. Figure 6.4 shows the cumulative frequency curve constructed from Table 6.9. The cumulative frequency curve usually has a characteristic S-shape

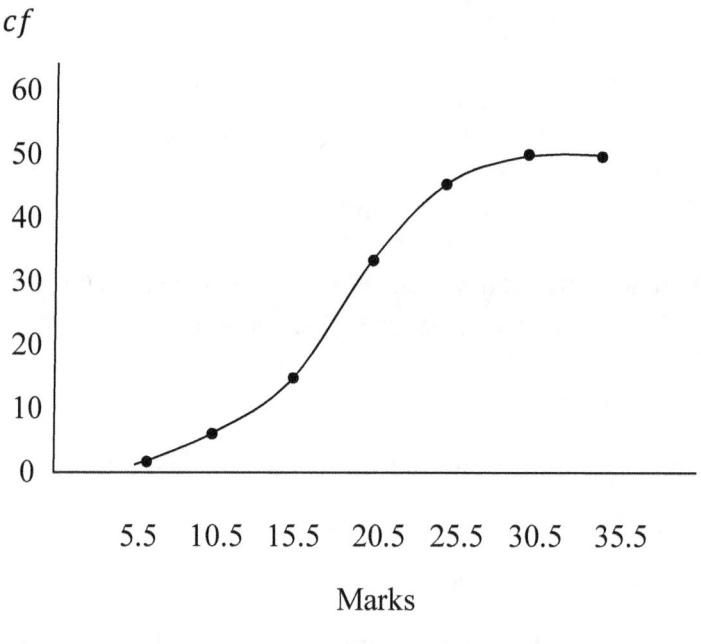

Figure 6.4

Try this 6

The marks obtained by 100 students in a test are shown in the table below

Marks	0-9	10 - 19	20 - 29	30 - 39	40 - 49	50 – 59	60 – 69	70 - 79	80 - 89	90 - 99
Frequency	3	7	8	12	17	20	14	10	5	4

Draw a cumulative frequency curve for the distribution

Exercise 6.3

1. The marks obtained by 100 students in a test are shown in the table below

Marks(%)	0 - 9	10 - 19	20 - 29	30 - 39	40 – 49	50 - 59	60 - 69	70 – 79	80 - 89	90 - 99
Frequency	3	5	8	17	23	20	12	7	3	2

(b) Draw a cumulative frequency curve

(b) How many students obtained at least 65%?

2. The following shows the frequency distribution of marks obtained by 200 students.

Marks	0 - 9	10 - 19	20 - 29	30 - 39	40 - 49	50 - 59	60 - 69	70 - 79	80 - 89	90 - 99
Frequency	6	10	14	34	48	40	26	12	8	2

(a) Draw a cumulative frequency curve

(b) Find the pass mark if 25% of the candidates failed

3. The ages in years of 50 people at a party are shown in the table

Ages	6 - 10	11 - 15	16 - 20	21 - 25	26 - 30	31 - 35	36 - 40	41 - 45	46 - 50
Frequency	2	4	15	8	9	6	3	2	1

(a) Draw a cumulative frequency curve

(b) How many people were at most 18 years old?

6.4 Measures of Central Tendency

When dealing with a set of numerical data, we often need to find a central value. There are three common measures of central values, the arithmetic mean, the median and the mode.

Arithmetic Mean

The arithmetic mean or simply the mean of n numbers is the sum of the numbers divided by n. For example, the arithmetic mean $(\bar{x})$ of the set $x_1,\ x_2,\ x_3,\ \cdots,x_n$ is

$$\bar{x} = \frac{x_1 + x_2 + x_3 + \cdots + x_n}{n}$$

written briefly as

$$\bar{x} = \frac{\sum x}{n}$$

where $\sum x$ = sum of all values

$\qquad n$ = number of values

Example

Find the mean of the five numbers 9, 4, 5, 12 and 10.

$$mean = \frac{9 + 4 + 5 + 12 + 10}{5}$$

$$= 8$$

Try this 7

Find the mean of the set of numbers 6, 4, 7, 10 and 3

If the numbers $x_1, x_2, x_3, \cdots, x_n$ occur with frequencies $f_1, f_2, f_3, \cdots, f_n$ respectively the arithmetic mean can be computed as

$$\bar{x} = \frac{f_1 x_1 + f_2 x_2 + f_3 x_3 + \cdots + f_n x_n}{f_1 + f_2 + f_3 + \cdots + f_n}$$

In brief, we write

$$\bar{x} = \frac{\sum f x}{\sum f}$$

where $\sum fx$ = the sum of the product of the values and frequencies

   $\sum f$ = sum of frequencies

Example

The numbers 5, 8, 6 and 3 occur with frequencies 3, 2, 4 and 1 respectively. Find the arithmetic mean

$$mean = \frac{3(5) + 2(8) + 4(6) + 1(3)}{3 + 2 + 4 + 1}$$

$$= \frac{58}{10}$$

$$= 5.8$$

Try this 8

The numbers 2, 4, 12, 7, 8 and 9 occur with frequencies 4, 3, 1, 6, 2 and 4 respectively. Find the mean

Calculating the arithmetic mean from tabulated data

Ungrouped Data

Example

The table below shows the marks obtained in a test marked out of 5. Find the mean mark.

Marks	0	1	2	3	4	5
Frequency	6	7	9	6	6	4

It is often better to do the calculations in a table as shown below. The third column shows the values of the product of frequency and the corresponding mark.

Marks(x)	Frequency(f)	fx
0	6	0
1	7	7
2	9	18
3	8	24
4	6	24
5	4	20
	$\Sigma f = 40$	$\Sigma fx = 93$

Table 6.10

$$mean = \frac{\Sigma fx}{\Sigma f}$$

$$= \frac{93}{40}$$

$$= 2.325$$

Try this 9

The ages of children at a party is shown in the table below. Find the mean age.

Ages(in years)	5	6	7	8	9	10
Frequency	4	5	11	6	3	1

Grouped Data

To compute the arithmetic mean of a grouped data, we interpret x as the class midpoint.

Example

The table below shows the ages of a group of pupil at a concert. Find the mean age.

Ages(in years)	11 - 15	16 - 20	21 - 25	26 – 30	31 – 35
Frequency	6	3	4	5	2

Begin by constructing a table like that shown below. To find the class midpoints, add the class limits and divide by 2.

Ages	Frequency(f)	Class midpoint(x)	fx
11 – 15	6	13	78
16 – 20	3	18	54
21 – 25	4	23	92
26 – 30	5	28	140
31 – 35	2	33	66
	$\Sigma f = 20$		$\Sigma fx = 430$

Table 6.11

$$mean = \frac{\Sigma fx}{\Sigma f}$$

$$= \frac{430}{20}$$

$$= 21.5$$

Try this 10

The marks of students marked out of 50, is shown in the table below. Find the mean mark.

Marks	21 – 25	26 – 30	31 - 35	36 – 40	41 – 45	46 – 50
Frequency	4	3	5	4	3	1

Exercise 6.4(a)

1. Find the mean of the following data:

(a) 13, 15, 16, 17, 18

(b) 41, 51, 63, 70, 55

(c) 16, 15, 17, 18, 16, 19, 24, 25, 22

(d) 30, 31, 35, 20, 26, 22, 32, 23, 25, 26

2. The numbers 24, 25, 27, 28 and 30 occur with frequencies 3, 4, 6, 5 and 2 respectively. Find the mean.

3. The numbers 12, 13, 15, 16, 19 and 20 occur with frequencies 2, 3, 5, 6, 3 and 1 respectively. Find the mean.

4. Find the mean of the following data:

(a) 2, 5, 4, 4, 5, 7, 5, 9, 8, 7, 7, 9, 8, 7, 8

(b) 31, 32, 35, 35, 35, 38, 44, 44, 45, 47

5. Find x if the mean of the following set of values 6, 8, x, 7, 14, 16 is 10

6. Find x, if the mean of the following set of values 11, 13, x, 17, 18 is 14.8

7. 6, 8, 9, x, 11, 13 occur with frequencies 2, 3, 4, 6, 3 and 2 respectively. Find x, if the mean is 8.65

8. The table below gives the frequency distribution of marks scored by a class in a test.

Marks	3	4	5	6	7	8	9	10
Frequency	1	2	3	6	7	3	2	1

Find the mean mark

9. The table below gives the ages of students in a school football team

Ages(in years)	14	15	16	17	18	19
Frequency	2	3	4	6	4	1

Calculate the mean age

10. Find the mean mark of the distribution in the table below

Mark	8 – 10	11 – 13	14 – 16	17 – 19	20 – 22	23 – 25
Frequency	5	10	9	16	6	4

11. The table gives the ages (in years) of a group of people in a club

Ages	7 – 10	11 – 14	15 – 18	19 – 22	23 – 26	27 – 30
Frequency	2	3	7	9	5	4

Find the mean age, correct to the nearest year

12. The table below shows the distribution of weights (in kg.) of students in a survey

Weight	1 - 10	11 – 20	21 – 30	31 – 40	41 – 50	51 – 60	61 – 70
Frequency	3	7	12	14	7	5	2

Find the mean weight

Mode

The mode of a set of numbers is the number that occur most frequently in the list

For example, the mode of the numbers 11, 12, 16, 14, 12, 15 is 12. Notice that 12 occur twice.

A set of numbers may have more than one mode. For example, the set of numbers:

11, 12, 13, 13, 15, 17, 20, 13 20, 15, 20 has two modes 13 and 20. Notice that both 13 and 20 occur three times.

The set of values has no mode if each data value has a frequency of 1 or if the frequencies of all the data values are the same. For example, the set of numbers 2, 2, 5, 5, 3, 3, 1, 1 has no mode

Try this 11

Find the mode of the following set of numbers:

(a) 12, 16, 16, 15, 14, 16, 17

(b) 2, 4, 5, 6, 5, 8, 9, 10, 4

Modal Class

The class interval with the largest frequency is called the modal class

Exercise 6.4(b)

1. Find the mode of the following set of numbers:

(a) 5, 6, 7, 6, 5, 8, 5, 8, 4, 5

(b) 12, 14, 15, 16, 15, 18, 19, 20, 14

(c) 0, 1, 3, 2, 3, 0, 3

(d) 25, 26, 24, 26, 21, 26, 20, 26, 21, 23, 21, 27, 21

(e) 1, 5, 8, 3, 2, 6

2. The table shows the marks obtained by 20 students in a test marked out of 5

Marks	1	2	3	4	5
Frequency	4	3	5	6	2

Find the modal mark

3. Find the mode of the distribution below

x	5	6	7	8	9	10	11	12	13	14	15
f	6	8	7	9	10	11	15	12	11	7	7

4. State the modal class for the distribution below

Marks	10 – 14	15 – 19	20 – 24	25 – 29	30 – 35
Frequency	7	8	11	6	8

5. The masses of 100 parcels at a post office is shown in the table below

Mass(kg)	1 – 5	6 – 10	11 – 15	16 – 20	21 – 25
Frequency	15	17	26	24	18

State the modal class

Median

The median of a set of numbers arranged in order is the middle value, if the number of values is odd or the mean of the two middle values, if the number of values is even.

For example, the median of the set of numbers 2, 3, 5, 6, 8 is 5.

The median of the set of numbers 11, 12, 13, 14, 15, 16, 17, 18 is the mean of the two middle numbers, i.e. the median is $\frac{14+15}{2} = 14.5$

Try this 12

Find the median for the set of numbers:

(a) 0, 1, 2, 3, 4, 5, 6   (b) 5, 6, 8, 9, 10, 12   (c) 3, 6, 7, 2, 4, 8, 3

Median Position

If n numbers are arranged in order the median is the value in the $\frac{n+1}{2}$ th position. For example, the median of the set of numbers: 6, 7, 8, 9, 11, 12, 13, 14 and 15 is the value in the $\frac{9+1}{2}$ i.e. 5 th position. For the set of numbers: 4, 5, 5, 7, 9, 11, 12 and 15 the median position is $\frac{8+1}{2}$ i.e. $4\frac{1}{2}$ position. Since the $4\frac{1}{2}$ position is mid-way between the

4 th and 5 th position, the mean of the numbers in the 4 th and 5 th position gives the median.

Try this 13

Find the median position of the set of values below:

(a) 0, 1, 3, 5, 5, 7, 8   (b) 5, 6, 7, 8, 10, 13, 16, 17, 18, 19

Finding the median from a frequency distribution table

Example

The table below shows the height (in mm) of 20 plants in a nursery

Heights(in mm)	15	16	17	18	19	20
Frequency	1	2	3	6	5	3

Find the median height

Begin by adding a row for the cumulative frequency.

Heights	15	16	17	18	19	20
Frequency	1	2	3	6	5	3
Cf	1	3	6	12	17	20

The median value is the $\frac{20+1}{2}$ i.e. $10\frac{1}{2}$ th value. By looking at the cumulative frequency row, you can see that both the 10 th and 11 th values are 18. Therefore the median height is $\frac{18+18}{2} = 18$.

Try this 14

Find the median for the distribution below:

x	13	14	15	16	17	18
f	2	3	4	6	3	2

Quartiles, Deciles and Percentiles

The median separates a data arranged in order into two equal parts. If each part is subdivided by a new median the result is four equal parts. Each of the three separating values is called a quartile. These values are denoted by Q_1, Q_2 and Q_3. Q_1 and Q_3 are called first (or lower) quartile and third (or upper) quartile respectively. Q_2 is the median.

The values which divide a data into ten equal parts are called the deciles while the values which divides a data into one hundred equal parts are called percentiles.

Finding the Quartiles

To find the quartiles, first divide the data values into two equal parts, ignoring the middle value when the number of values is odd.

Example

Find the first and third quartiles for the set of values:

$$1, 2, 3, 5, 7, 8, 10, 13, 15, 17$$

Example

First, divide the data into two groups of 5 numbers each. The two groups are:

1, 2, 3, 5, 7 and 8, 10, 13, 15, 17

The median of the first half is 3, so the lower quartile is 3, and the median of the upper half is 13, so the upper quartile is 13.

Try this 15

Find the first and third quartiles for the set of values:

$$15, 16, 18, 18, 20, 22, 25, 27, 29, 31, 32, 35$$

Exercise 6.4(c)

1. Find the median of the set of values below:

(a) 3, 5, 6, 7, 8   (b) 15, 16, 17, 18, 20, 22, 23, 24

(c) 1, 2, 3, 4, 6, 7, 9   (d) 1, 2, 5, 5, 6, 7, 8, 9, 9, 11

2. Find the median of the following set of data:

(a) 17, 15, 13, 18, 16

(b) 30, 32, 27, 25, 26, 34, 33, 28

(c) 14, 12, 13, 12, 17, 16, 11

(d) 11, 26, 9, 15, 21, 17, 25, 9, 7, 8, 12

3. The table below shows the ages of students in a class

Ages(in years)	17	18	19	20	21
Frequency	2	5	7	4	2

Find the median age

4. The table shows the marks obtained by a group of students in a test

Marks	0	1	2	3	4	5
Frequency	3	7	8	4	5	3

Find the median mark

5. Find the lower quartile and the upper quartile of the following data:

(a) 2, 3, 4, 5, 6, 8, 10, 12, 15, 16, 18

(b) 20, 21, 24, 26, 27, 29, 30, 31, 33, 34, 35, 36, 40, 42, 45

(c) 11, 26, 9, 15, 21, 17, 25, 9, 7, 8, 12

Using Statistical Graphs

Estimating the mode from a Histogram

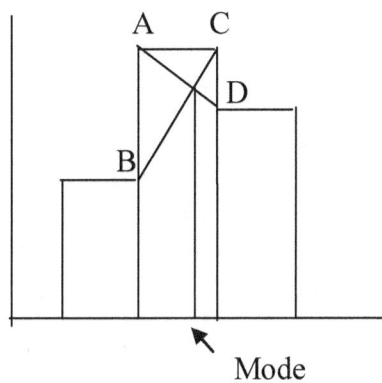

Figure 6.5

Figure 6.5 shows three bars of the histogram of a frequency distribution. The modal class is the middle class. The value of the mode lies somewhere within this range and its value can be found by the construction shown. The value on the horizontal axis of the point of intersection of the diagonal lines AD and BC is taken as the mode of the distribution.

Example

The masses of 40 castings gave the following frequency distribution

Mass(kg)	10 - 12	13 - 15	16 - 18	19 - 21	22 - 24
Frequency	3	6	15	9	7

Draw a histogram and use it to estimate the mode.

Figure 6.6 shows the histogram of the distribution

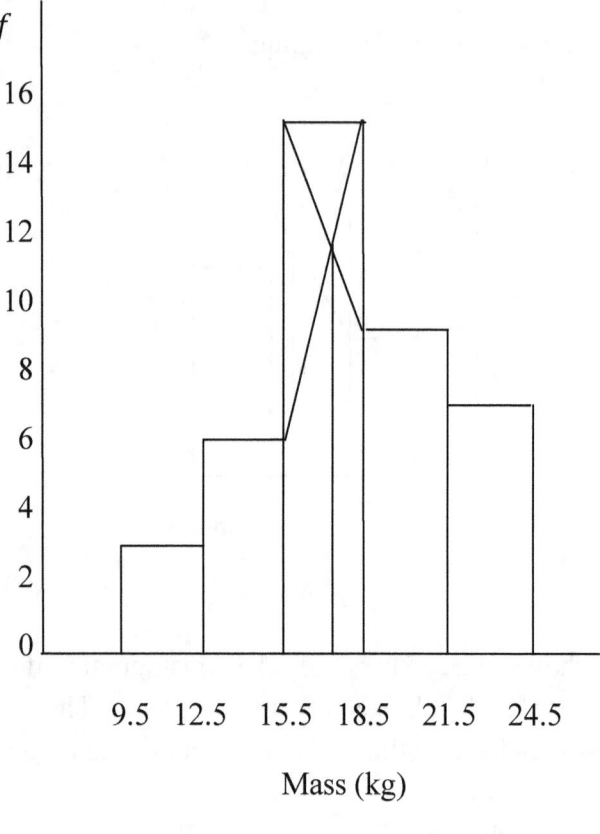

Mass (kg)

Figure 6.6

The modal class is the third class with boundaries 15.5 kg and 18.5 kg. The mode is 17.3 kg.

Try this 16

The table below gives the height of a group of children at a nursery

Height(cm)	50 - 54	55 - 59	60 - 64	65 - 69	70 - 74	75 - 79	80 - 84
Frequency	3	7	10	5	8	2	1

Draw a histogram and use it to estimate the modal height.

Using the cumulative frequency curve

We can estimate the median, the quartiles and percentiles from a cumulative frequency curve.

Example

The frequency distribution shows the masses of 40 castings.

Mass(kg)	10 - 12	13 - 15	16 - 18	19 - 21	22 - 24	25 - 27	28 - 30
Frequency	3	5	9	12	5	4	2

Draw a cumulative frequency curve and use it to find the median, the first and third quartiles.

Before drawing the cumulative frequency curve you need to construct a cumulative frequency table.

Mass less than	Cumulative frequency
12.5	3
15.5	8
18.5	17
21.5	29
24.5	34
27.5	38
30.5	40

The cumulative frequencies are plotted against the upper class boundaries. Cumulative frequency is plotted on the vertical axis. The curve is shown in Figure 6.6.

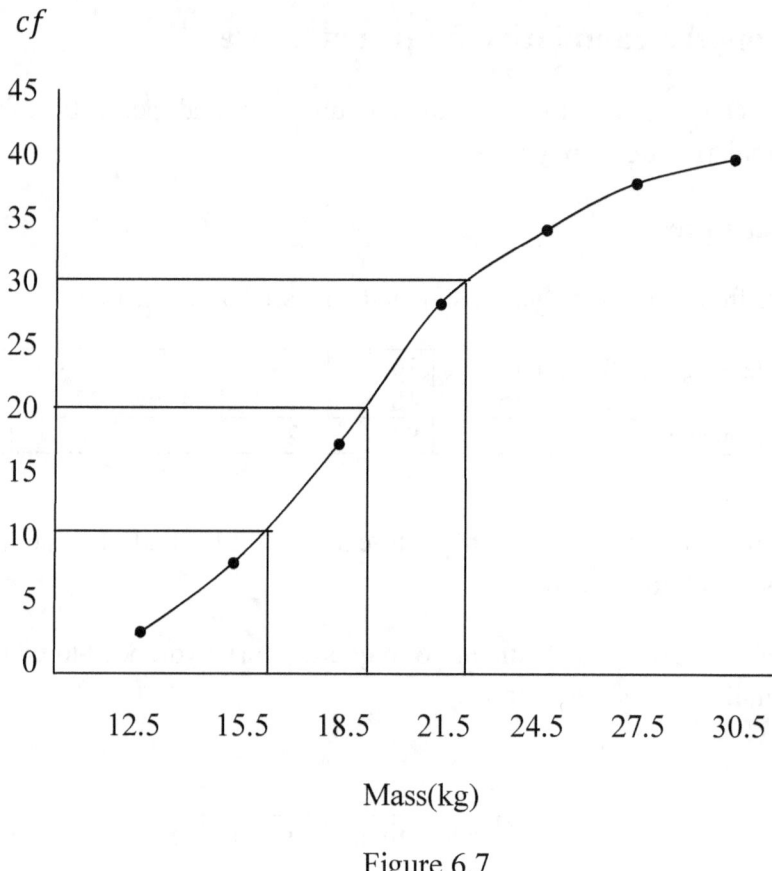

Figure 6.7

The total frequency is 40. The median is the 20th value. The median is the value on the horizontal axis corresponding to the cumulative frequency value of 20, as shown in Figure 6.7. The median is 19.1 kg

The first and third quartiles correspond to the $\frac{1}{4} \times 40$ i.e. 10th value and $\frac{3}{4} \times 40$ i.e. 30th value respectively. The first and the third quartiles are 16.7 kg and 21.8 kg respectively.

Try this 17

The table below gives the distribution of the masses of 80 male students recorded to nearest kg. Draw a cumulative frequency curve and use it to find the median, the first and third quartiles.

Mass(kg)	50 - 54	55 - 59	60 - 64	65 - 69	70 - 74	75 - 79	80 - 84	85 - 89
Frequency	2	5	9	18	23	14	8	1

Exercise 6.4(d)

1. The table below shows the distribution of the monthly (in GH¢)
income of 50 workers in a company

Monthly Income	120 - 129	130 - 139	140 - 149	150 - 159	160 - 169
Frequency	5	9	16	14	6

Draw a histogram for the distribution and use it to estimate the
modal income.

2. The distribution of ages of 30 school children is shown in the table
below

Age(yrs)	12 - 14	15 - 17	18 - 20	21 - 23	24 - 26
Frequency	5	7	10	6	2

Draw a histogram for the distribution and use it to estimate the
modal age.

3. The table shows the mark distribution of a class of 25 students

Marks(%)	0 - 9	10 - 19	20 - 29	30 - 39	40 - 49	50 - 59	60 - 69
Frequency	4	5	2	3	6	3	2

Draw a histogram for the distribution and use it to estimate the modal
mark.

4. The table below gives the distribution of the marks obtained by 100
students in a test

Marks	0- 9	10 - 19	20 - 29	30 - 39	40 - 49	50 - 59	60 - 69	70 - 79	80 - 89	90 - 99
Frequency	4	6	10	10	13	17	20	15	4	1

Draw a cumulative frequency curve for the distribution and use it to estimate the median, the first and third quartiles

5. The table below shows the marks scored by 100 candidates in a test

Marks(%)	1 - 10	11 - 20	21 - 30	31 - 40	41 - 50	51 - 60	61 - 70	71 - 80	81 - 90	91 - 100
Frequency	4	6	10	20	30	12	8	6	3	1

Draw a cumulative frequency curve for the distribution and use it to estimate the median mark, the first and third quartile

6. The table below gives the distribution in centimetres, of the lengths (in cm.) of 100 wooden pegs

Length	10 - 19	20 - 29	30 - 39	40 - 49	50 - 59	60 - 69	70 - 79	80 - 89	90 - 99
Frequency	5	7	12	19	23	14	10	9	1

Draw the cumulative frequency curve for the distribution and use it to estimate the 60 th percentile. How many pegs were at least 75 centimetres long?

7. The table gives the scores of 50 students in a test

Scores	5 - 9	10 - 14	15 - 19	20 - 24	25 - 29	30 - 34	35 - 39
No. Of students	4	5	12	16	7	4	2

Draw the cumulative frequency curve for the distribution and use it to estimate the pass mark if 80% of the students passed.

8. The table below shows the distribution of mass (in kg.) of students in a survey

Mass(kg)	20 - 24	25 - 29	30 - 34	35 - 39	40 - 44	45 - 49	50 - 54	55 - 59
Frequency	2	3	7	26	29	25	6	2

Draw a cumulative frequency curve of the distribution and use it to estimate the 90th percentile and the interquartile range

6.5 Measure of Dispersion

The table below shows the ages of two groups of 5 students in a class

Group A	10	11	12	13	14
Group B	8	9	12	14	17

The mean ages for both group is 12. In group A there is a close clustering of the scores about the mean. On the other hand, in group B the scores are widely scattered about the mean. It is therefore very important to have a measure of the spread (or dispersion) of a distribution as well as its mean. Some measures of dispersion are the range, the interquartile range, semi-interquartile range, mean deviation and standard deviation.

Range

The difference between the largest and smallest numbers in a data is called the range of the data.

Example

Determine the range of the following set of values 9, 12, 13, 14, 15, 17

The largest number is 17 and the smallest number is 9. So the range is $17 - 9 = 8$

Try this 18

Determine the range of the following set of values 25, 14, 11, 26, 28, 31, 13, 30, 45

Interquartile Range

The difference between the upper quartile and the lower quartile is called the interquartile range.

Interquartile range = upper quartile − lower quartile

Example

Find the interquartile range for the following set of data:

5, 6, 7, 10, 12, 13, 15, 18, 19, 20, 25

The lower and upper quartiles are 7 and 19 respectively. Therefore the interquartile range is $19 - 7 = 12$

Try this 19

Find the interquartile range for the following set of data:

11, 12, 15, 17, 18, 20, 22, 25, 27, 30, 33, 35, 38, 40, 49

Semi-interquartile Range (Quartile deviation)

The semi − interquartile range of a set of data is define by

$$Semi - interquartile\ range = \frac{upper\ quartile - lower\ quartile}{2}$$

Example

Find the semi–interquartile range of the following set of data:

12, 13, 15, 16, 17, 18, 19, 20, 27, 28, 30

The lower quartile and the upper quartile are 15 and 27 respectively. Therefore,

$$Semi - interquartile\ range = \frac{27 - 15}{2} = 6$$

Try this 20

Find the semi- interquartile range of the following set of data:

21, 42, 53, 55, 56, 57, 58, 59, 60, 67, 69, 70, 75

Mean Deviation

The mean deviation is the arithmetic mean of the absolute differences of each value from the mean. The mean deviation for a data of n numbers is obtained as follows:

1. First find the mean of the numbers

2. Next, subtract the mean from each number, and then add the absolute value of each difference.

3. Finally divide the sum by the number of values.

For instance, the mean deviation of a set of n numbers x_1, x_2, x_3. $\cdots$, x_n is given by

$$mean\ deviation = \frac{\Sigma|x - \bar{x}|}{n}$$

where $\bar{x}$ is the arithmetic mean of the numbers.

If x_1, x_2, x_3. $\cdots$, x_n occur with frequencies f_1, f_2, f_3, $\cdots$ f_n respectively, the mean deviation is given by

$$mean\ deviation = \frac{\Sigma f|x - \bar{x}|}{\Sigma f}$$

This form is useful for grouped data where x represents the class midpoint, and f is the corresponding class frequency.

Examples

Find the mean deviation of the set of numbers 7, 8, 11, 13 and 16

$$arithmetic\ mean = \frac{7 + 8 + 11 + 13 + 16}{5}$$

$$= 11$$

$$mean\ deviation = \frac{|7 - 11| + |8 - 11| + |11 - 11| + |13 - 11| + |16 - 11|}{5}$$

$$= \frac{|-4| + |-3| + |0| + |2| + |5|}{5}$$

$$= \frac{14}{5}$$

$$= 2.8$$

The mean deviation is 2.8

Try this 21

Find the mean deviation of the set of numbers 12, 15, 17, 13, 16, 17

The marks obtained by 20 students in a mathematics test are shown by the distribution below:

Marks	15 - 19	20 - 24	25 – 29	30 - 34	35 – 39
Frequency	3	6	5	4	2

Find the mean deviation

The work is arranged as in the Table below:

Marks	f	Class midpoint(x)	fx	$\lvert x - \bar{x} \rvert$	$f\lvert x - \bar{x} \rvert$
15 – 19	3	17	51	9	27
20 – 24	6	22	132	4	24
25 – 29	5	27	135	1	5
30 – 34	4	32	128	6	24
35 – 39	2	37	74	11	22
	Σf $= 20$		$\Sigma fx = 520$		$\Sigma\lvert x - \bar{x} \rvert$ $= 102$

$$mean = \frac{\Sigma fx}{\Sigma f}$$

$$= \frac{520}{20}$$

$$= 26$$

$$mean\ deviation = \frac{\Sigma f\lvert x - \bar{x} \rvert}{\Sigma f}$$

$$= \frac{102}{20}$$

$$= 5.1$$

Try this 22

Find the mean deviation for the distribution below:

x	10	11	12	13	14	15	16
y	5	4	2	3	2	6	3

Standard Deviation

The standard deviation is used to describe the extent to which a collection of data spread around its mean. A small standard deviation means that the values tend to be close to their mean. A large standard deviation means that the values are widely scattered about their mean.

The standard deviation for a data of n values is obtained as follows:

1. first calculate the arithmetic mean

2. find the difference between the mean and each value

3. square each of the differences

4. sum the squared values

5. divide the sum by n, and finally

6. take the non negative square root of the quotient

In brief, the standard deviation for the n values $x_1, x_2, x_3, \cdots, x_n$ is given by

$$s = \sqrt{\frac{\Sigma(x - \bar{x})^2}{n}}$$

where s is the standard deviation

$\quad\quad$ $\bar{x}$ is the arithmetic mean

and n is the number of values

Examples

Determine the standard deviation of the set of values 2, 4, 6, 8, 10

The arithmetic mean

$$\bar{x} = \frac{\Sigma x}{n}$$

$$= \frac{2 + 4 + 6 + 8 + 10}{5}$$

$$= 6$$

x	$x - 6$	$(x - 6)^2$
2	- 4	16
4	- 2	4
6	0	0
8	2	4
10	4	16
		40

$$s = \sqrt{\frac{40}{5}}$$

$$= 2.8$$

Note that the standard deviation cannot be negative, and when two sets of data are compared, the one with the larger dispersion will have the larger standard deviation.

Try this 23

Determine the standard deviation of the set of values 6, 9, 9, 11, 15, 16

For a frequency distribution, the standard deviation is given by

$$s = \sqrt{\frac{\Sigma f(x - \bar{x})^2}{\Sigma f}}$$

The set of values 5, 6, 8, 9 and 14 occur with frequencies 1, 3, 2, 3 and 1 respectively. Determine the standard deviation.

The arithmetic mean

$$\bar{x} = \frac{\Sigma fx}{\Sigma f}$$

$$= \frac{1(5)+3(6)+2(8)+3(9)+1(14)}{1+3+2+3+1}$$

$$= \frac{80}{10}$$

$$= 8$$

x	f	$x - \bar{x}$	$(x - \bar{x})^2$	$f(x - \bar{x})^2$
5	1	- 3	9	9
6	3	-2	4	12
8	2	0	0	0
9	3	1	1	3
14	1	6	36	36
	10			60

$$s = \sqrt{\frac{60}{10}}$$

$$= 2.45$$

Try this 24

The set of values 11, 12, 14, 15 and 18 occur with frequencies 1, 2, 3, 3 and 1 respectively. Determine the standard deviation.

Variance

The variance of the set of values is the square of the standard deviation. The variance of the set of values $x_1, x_2, x_3, \cdots, x_n$ is found from the formula

$$varience = \frac{\Sigma(x - \bar{x})^2}{n}$$

This is equivalent to:

$$varience = \frac{\Sigma x^2}{n} - \bar{x}^2$$

Example

Find the variance of the set of values 1, 3, 5, 6 and 9

$$\bar{x} = \frac{1 + 3 + 5 + 6 + 9}{5}$$

$$= 4.8$$

Using the formula

$$varience = \frac{\Sigma x^2}{n} - \bar{x}^2$$

we have

$$variance = \frac{1^2 + 3^2 + 5^2 + 6^2 + 9^2}{5} - 4.8^2$$

$$= 7.36$$

Try this 25

Find the variance of the set of values 5, 6, 8, 9 and 12

Exercise 6.5

1. Find the range of the following sets of values:

(a) 8, 10, 12, 13, 15, 17   (b) 41, 43, 38, 47, 39, 50

2. Find the interquartile range of the following data

(a) 4, 7, 8, 9, 9, 11, 13, 16, 19, 20

(b) 12, 9, 7, 5, 15, 11, 13, 6, 5, 3, 17, 18

3. Find the semi-interquartile range of the following data

(a) 15, 16, 18, 19, 21, 23, 25

(b) 20, 22, 25, 25, 27, 28, 29, 32, 34, 36, 38

4. The table gives the scores of 50 students in a test

Scores	5 - 9	10 - 14	15 - 19	20 24	25 - 29	30 - 34	35 - 39
No. of students	4	5	12	16	7	4	2

Draw the cumulative frequency curve for the distribution and use it to estimate the semi-interquartile range.

5. The table below shows the distribution of weights of students in a survey

Mass(kg)	20 - 24	25 - 29	30 - 34	35 - 39	40 - 44	45 - 49	50 - 54	55 - 59
Frequency	2	3	7	26	29	25	6	2

 Draw a cumulative frequency curve of the distribution and use it to estimate the interquartile range

6. Find the mean deviation of each set of values

(a) 8, 12, 14 and 20   (b) 14, 6, 28, 31 and 21

(c) 7, 5, 10, 6, 9, 8, 14, 6, 5 and 4   (d) 10.5, 6.2, 8.6, 11.5 and 9.2

7 Find the mean deviation for the distribution below:

x	31	33	35	37	38	40	42
f	4	5	2	2	1	9	2

8. The heights of a collection of 50 plants were as given by the table below. Find the mean deviation of the heights

Heights (in cm)	0 - 4	5 - 9	10 - 14	15 - 19	20 - 24
Frequency	14	10	8	11	7

9. The following distribution shows the performance of 50 students in a test. Find the mean deviation of the distribution.

Marks	15 - 19	20 - 24	25 - 29	30 - 34	35 - 39
Frequency	8	14	12	9	7

10. The table shows the distribution of outcomes when a die is thrown 100 times. Calculate the mean deviation of the distribution.

Scores	1	2	3	4	5	6
Frequency	4	10	26	22	18	20

11. Find the standard deviation of the following set of data

(a) 11, 13, 15, 16, 19   (b) 7, 8, 9, 11, 12, 15, 17, 21, 25

12. The set of values 12, 13, 14, 15, 16, and 17 occur with frequencies 3, 4, 5, 4, 3 and 1 respectively. Find the standard deviation.

13. Find the variance of the following set of data

(a) 42, 38, 50, 50, 55   (b) 75, 65, 52, 68, 90

(c) 70, 38, 50, 30, 22, 44, 52, 21, 76, 45

6.6 Scatter Diagrams

The table shows how many days in a week eight students review the previous work and the marks obtained in a mathematics test.

Number of Days	1	2	3	2	4	5	4	5
Marks	32	48	65	62	72	76	82	90

These results can be plotted on a graph, taking the number of days as x- coordinate of the point and the marks as the y- coordinate. This gives the graph shown below.

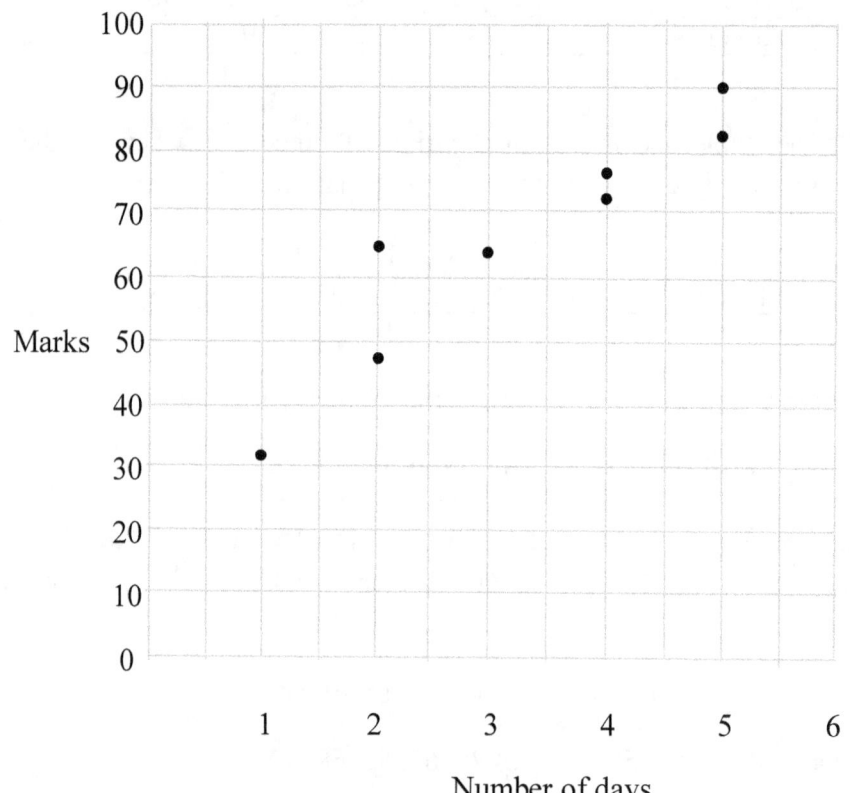

Number of days

This graph is called a scatter diagram (or a scatter plot). Scatter diagrams help us see the relationship between two sets of data. For example, you can see from the graph that spending more days to review previous work leads to higher marks. We say that there is correlation between the two sets of data.

Types of Correlation

The word correlation is used to describe how closely the points of a scatter diagram are to a straight line. The relationship between two sets of data can be described as positive, negative or no correlation.

Positive Correlation

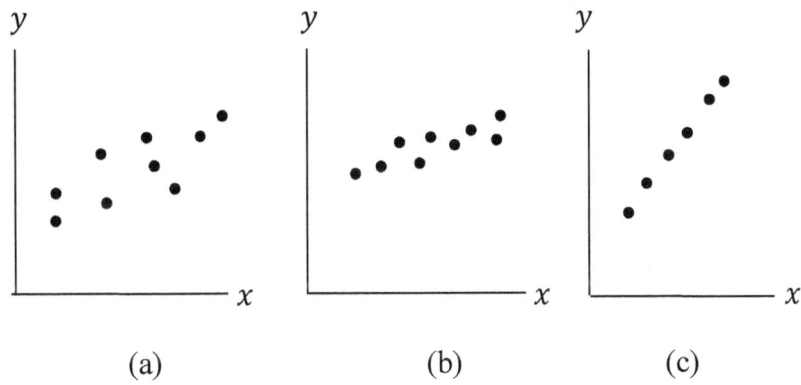

(a)   (b)   (c)

In each of the three cases, you can see that as x values increase, the y values tend to increase. We say that there is positive correlation between the data.

A pattern of points which tends to form a straight line suggests a high correlation between the set of data. The correlation is low in case (a) and high in case (b). In case (c) we have a perfect positive correlation.

Negative Correlation

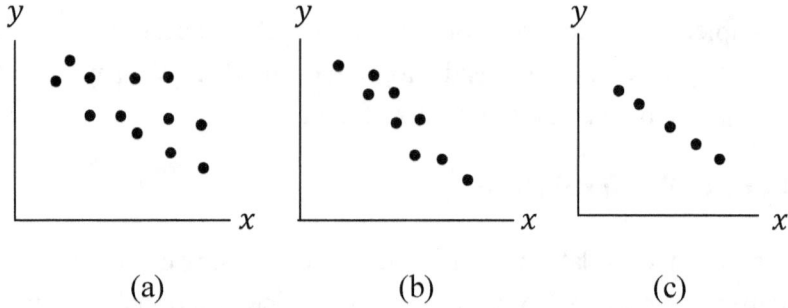

(a)   (b)   (c)

In each of the cases, you can see that as the x values increase, the y values tend to decrease. We say that there is negative correlation between the data. We have a low negative correlation in case (a) and high negative correlation in case (b). In case (c) we have a perfect negative correlation.

No Correlation

A set of data can give a scatter diagram as shown below

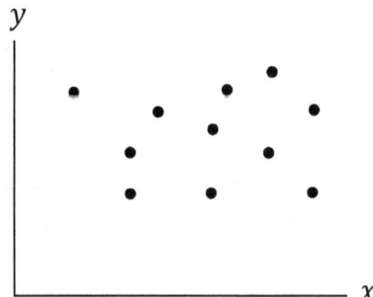

The points are scatted at random, and do not seem to cluster around a line. There is no obvious relationship between x and y. We say that there is no correlation.

Drawing the line of best fit

Often the points on a scatter diagram do not lie on a straight line. We could draw a straight line which comes closer to fitting all points, called the line of best fit.

The position of the line of best can be estimated by eye. The line must be drawn through as many points as possible and must be close to an equal number of points on each side.

Example

The table below shows the marks obtained by eight students in a test in Mathematics and Science.

Mathematics	2	3	4	5	6	8	10	12
Science	1	2	4	4	5	7	8	9

(a) Plot a scatter diagram of Mathematics marks against Science marks

(b) Describe the correlation between the two set of data

(c) Draw a line of best fit on your scatter diagram

(d) Use your line of best fit to estimate the possible mark obtain in science by a student who obtain 7 in mathematics

(a) The scatter diagram is as shown below.

Science

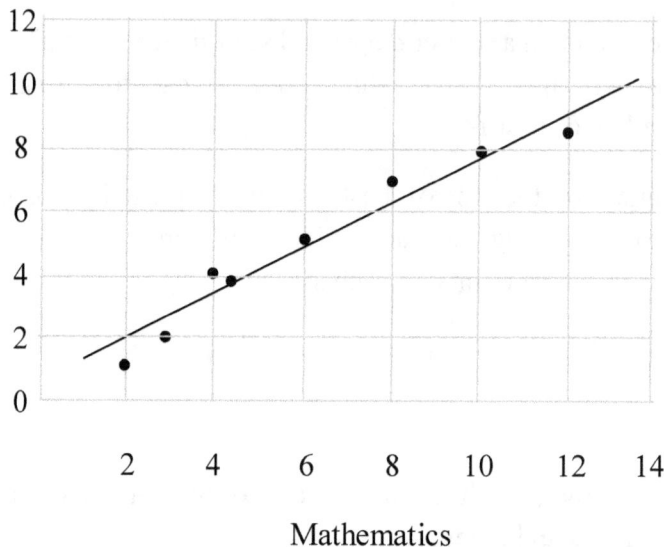

Mathematics

(b) There is a high positive correlation between the two variables

(c) The line of best fit is drawn to pass through the points as closely

as possible as shown

(d) The possible science mark is 5.6

Try this 26

Efua kept records of her weekly wages and the amount she spent on food. The table shows 9 different week's wages and the amount she spent on food that week.

Wages	75	180	125	137	93	45	110	176	23
Food Expenses	41	95	76	81	68	27	70	101	38

(a) Plot a scatter diagram of wages against food expenses

(b) Describe the correlation between the two variables

(c) Draw a line of best fit on your scatter diagram

(d) How much did she spend on food in a week her wage was

GH¢ 100?

Exercise 6.6

1. The table shows the amount of time 10 students spend studying
and the grade they get on a test

Hours Studied	16	25	34	18	11	9	27	40	15	17
Test Score	73	82	85	88	71	48	92	98	65	75

(a) Plot a scatter diagram of Hours Studied against Test Score

(b) Described the correlation between the two variables

(c) Draw a line of best fit on your scatter diagram

(d) What will be the test score for a student who studied 31 hours?

2. The amount of time 10 students spend studying and the grade they
get on a test are plotted on the scatter diagram

Test score

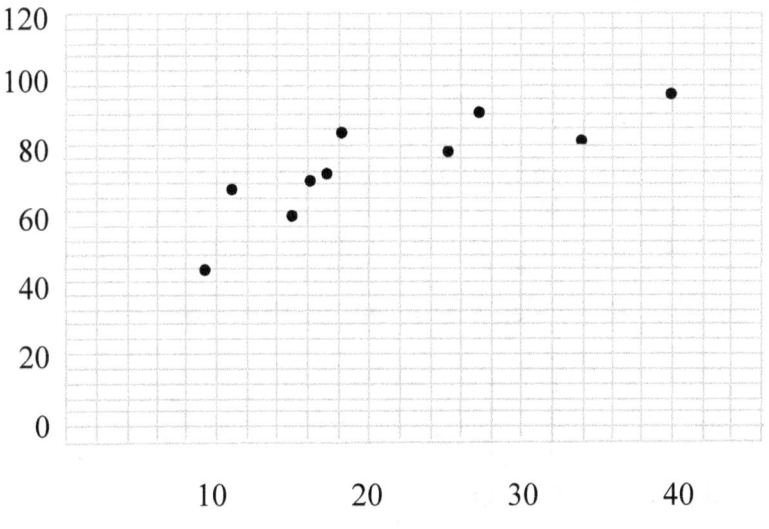

Hour studied

(a) Write down the test score of a student who study 34 hours.

(b) What type of correlation is shown?

3. In a factory, the number of work-hours in a safety training and the number of work-hour lost due to accidents have been recorded for 10 divisions.

Division	A	B	C	D	E	F	G	H	I	J
No. of Work-Hours in Safety Training	15	20	25	30	35	45	50	55	65	70
No. of Work-Hours lost from Accidents	85	80	77	75	65	58	55	53	47	40

(a) Plot a scatter diagram for the data

(b) Describe the correlation between the two variables

(c) Draw a line of best fit on your scatter diagram

(d) How many work-hours would be lost due to accident if 40 hours of safety training were given?

Review exercise 6

1. The ages, in years of 30 pupils in a survey are shown below

12	16	14	16	15	12
13	17	12	14	15	14
14	15	15	14	14	15
15	16	13	15	13	14
13	14	15	12	16	17

Form a frequency table for the distribution

2. The lengths of 40 wooden pegs in cm, are shown below

22	28	21	35	38	18	37	24
27	13	33	30	23	42	14	25
32	34	29	30	43	27	32	32
23	24	36	14	34	48	38	28
17	19	22	26	22	22	28	17

Form a grouped frequency table using the class intervals 11 – 15, 16 – 20, 21 – 25, etc.

3. The table below shows the number of pupils who offer certain subjects in a school.

History	Geography	Mathematics	Science	Others
25	15	35	20	5

Represent this information on a pie chart

4. The table gives the proportions in which Mr. Mensah spends his monthly salary

Food	Rent	Savings	Transport	Income Tax	Others
30%	15%	5%	7.5 %	20%	22.5%

Represent this information on a pie chart

5. Find the mean, mode and median for each of the following sets of values

(a) 4, 9, 11, 8, 6, 15, 3, 10, 9   (b) 16, 21, 11, 19, 22, 16, 15, 18

(c) 22, 23, 24, 26, 26, 29, 19, 26, 30, 27

6. The table shows the ages of 40 people in a club

Ages(in years)	1 - 10	11 - 20	21 - 30	31 - 40	41 - 50	51 - 60
Frequency	2	6	9	12	7	4

Draw a histogram and use it to estimate the modal age.

7. The table below shows the distribution of weights of students (in kg.) in a survey

Weight	20 - 29	30 - 39	40 - 49	50 - 59	60 - 69	70 - 79	80 - 89
Frequency	5	9	12	7	10	4	3

Draw a histogram and use it to estimate the modal weight.

8. The table bellow shows the distribution of marks obtained by students in a test marked out of 10

Marks	3	4	5	6	7	8	9	10
Frequency	3	5	8	11	13	5	3	2

Find the median mark

9. The table below gives the distribution of masses (kg) of youth at an interview

Mass(kg)	54	55	56	57	58	59	60
Frequency	6	7	4	5	8	6	4

Find the median mass

10. The table shows the distribution of marks obtained by some students in a test

Marks	13	14	15	16	17	18	19
Frequency	3	7	8	5	10	5	2

Calculate the mean mark of the distribution

11. The following table shows the distribution of marks obtained by 20 students in a test

Marks	10 - 14	15 - 19	20 - 24	25 - 29	30 - 35
Frequency	3	4	7	2	4

Calculate the mean mark

12. The table gives the ages of the members of a club

Ages(in years)	11 - 15	16 - 20	21 - 25	26 - 30	31 - 35
Frequency	6	4	5	3	2

Calculate the mean age

13. The frequency distribution for the masses in grams of 50 toys

Mass(in grams)	61 - 63	64 - 66	67 - 69	70 - 72	73 - 75	76 - 78	79 - 81
Frequency	6	10	18	28	22	12	4

Draw a cumulative frequency curve and use it to estimate the median, the first and third quartile

14. The marks obtained by students in a test are shown in the distribution below.

Marks	10 - 19	20 - 29	30 - 39	40 - 49	50 - 59	60 - 69	70 - 79	80 - 89	90 - 99
Frequency	2	4	10	11	10	6	4	2	1

Draw a cumulative frequency curve and use it to estimate the median, the first and third quartile

15. The frequency distribution below shows the lengths of leaves of a certain plant in millimetre

Length	20 - 24	25 - 29	30 - 34	35 - 39	40 - 44	45 - 49	50 - 54	55 - 59	60 - 64
Frequency	1	5	10	19	25	21	15	3	1

Draw the cumulative frequency curve and use it to estimate the median and the interquartile range

16. Find the range of the following sets of values

(a) 6, 8, 10, 12, 13, 15, 15   (b) 21, 23, 18, 27, 19, 30

(c) 13, 28, 11, 17, 23, 19, 27, 11, 9, 10, 14

17. Find the interquartile range for the following data

(a) 20, 21, 24, 26, 27, 29, 30, 31, 33, 34, 35, 36, 40, 42, 45

(b) 31, 46, 29, 35, 41, 37, 45, 29, 27, 28, 32

18. Find the semi- interquartile range for the following data

(a) 35, 36, 37, 38, 39, 40, 41, 43, 44, 45, 47, 48, 49, 50, 52

(b) 21, 36, 19, 25, 31, 27, 35, 19, 17, 18, 22

19. A student took 6 test in a class and had the following marks 85, 77, 87, 82, 89 and 90. Find the mean deviation for the marks

20. The following distribution shows the ages, in years, of 5 contestants at a beauty contest:17, 16, 18, 21and 23. Find the mean deviation of the ages

21. The table below gives the distribution of marks in a test. Find the mean deviation of the marks

Marks	5	6	7	8	9	10
Frequency	1	12	14	15	7	1

22. The following distribution shows the performance of 50 students in a test. Find the mean deviation of the distribution.

Marks	5 - 9	10 - 14	15 - 19	20 - 24	25 - 29
Frequency	5	10	17	11	7

23. Find the standard deviation of the following data

(a) 11, 12, 14, 15, 18   (b) 6, 8, 9, 9, 10, 11, 17, 20

(c) 37, 38, 39, 31, 32, 35, 37, 41, 45, 35

24. Find the variance of the following data

(a) 42, 38, 50, 50   (b) 15, 15, 12, 18, 20

(c) 25, 35, 45, 48, 50, 52, 52, 55

25. The table below gives the wages of 8 people and the number of days in a mouth they worked part time in a factory:

Number of days	10	11	12	13	14	15	16	17
Wages(in GH¢)	25	38	42	57	64	74	83	96

(a) Plot a scatter diagram of Number of days against Wages

(b) Describe the correlation between the two variables

(c) Draw a line of best fit on your scatter diagram

26. The table below gives the mid-term and final exam scores for 10 students.

Mid-Term Exam Scores	21	29	34	26	17	43	38	41	18	27
Final Exam Scores	62	67	70	63	57	72	69	78	64	65

(a) Plot a scatter diagram of the Mid-Term Exam Scores against Final Exam Scores

(b) Describe the correlation between the two variables

(c) Draw a line of best fit on your scatter diagram

(d) Predict the final exams scores of a student whose mid-term score was 34

Chapter Test 6

Take this test as you would take a test in a class. After you are done, check your work against the answers in the back of the book.

1. The data given below are marks obtained by 30 pupils in a test marked out of 10

5	6	8	7	6	7
8	7	5	8	7	8
7	8	6	10	8	9
9	10	7	9	5	6
6	8	6	8	9	8

(a) Form a frequency distribution table for the data

(b) State the modal mark

(c) Find the median mark

(d) Calculate the mean mark

2. The data below are heights in centimetres of a class of 30 pupils

153	157	162	168	163	163
167	163	152	162	178	172
156	151	143	166	152	173
167	166	172	157	172	178
163	148	163	161	153	168

(a) Form a frequency distribution, using class intervals

   140 – 144, 145 – 149, 150 – 154, etc.

(b) Construct a histogram for the distribution

(c) Estimate the modal height from the histogram

3. The table below shows the number of cars that passes a toll booth

in five months in 2009

January	February	March	April	May
140	110	190	520	240

   Draw a pie chart to show this result

4. The table gives the ages (in years) of a group of people in a club

Ages	7 - 10	11 - 14	15 - 18	19 - 22	23 - 26	27 - 30
Frequency	3	5	9	11	7	5

Calculate the mean age

5. The marks obtained by a group of students in a mathematics test

are shown in the distribution below.

Marks	10 - 19	20 - 29	30 - 39	40 - 49	50 - 59	60 - 69	70 - 79	80 - 89	90 - 99
Frequency	2	5	9	11	9	7	4	2	1

(a) Draw a cumulative frequency curve

(b) Use the cumulative frequency curve to estimate

(i) the median mark

(ii) the interquartile range

6. Find the range, and the semi-interquartile range for the data below

   17, 16, 18, 21, 22, 19, 21, 18, 22, 24, 23

7. The frequency distribution below shows the distribution of weekly incomes of 50 workers. Find the mean deviation of the weekly incomes

Weekly incomes (GH¢)	25 - 29	30 - 34	35 - 39	40 - 44	45 - 49
Number of workers	5	16	9	14	6

8. Calculate the standard deviation of the numbers

   22, 28, 34, 24, 36, 26, 30, 32

9. Calculate the variance of the numbers

   12, 7, 4, 16, 5, 12, 10, 9, 8, 7

10. The table below gives the number of hours 12 students spent studying for a test and how many incorrect answers they gave:

Number of Hours	5	6	9	8	5	4	4	10	7	9	2	8
Incorrect Answers	5	3	1	3	6	6	8	1	4	2	7	2

(a) Plot a scatter diagram of Number of Hours against Incorrect Answers

(b) Describe the correlation between the two variables

(c) Draw a line of best fit on your scatter diagram

(c) Using your line of best fit, predict the number of incorrect answers that a student who studied for 3 hours would give

7

Transformation

Recall that a mapping is a relation between two sets in which each member of the first set has an image in the second set. Both of the sets involved in a mapping can be points. A transformation maps an original figure, called the object (or the pre-mage) onto a final figure, called the image. There are various ways to transform an object. We will discuss four ways to transform an object.

Rigid Motion

An object undergoing a transformation changes in position, size or shape. A transformation which does not change the size or shape of an object is called a rigid motion. Three such transformations are Reflection, Rotation and Translation.

7.1 Reflection

You may be familiar with reflection in a plane mirror. The image in a plane mirror is as far behind the mirror as the object is in front of it. Also, the image is a lateral inversion of the object, that is, what is on the right of the object appears to be on the left in the image. In this section, we will discuss reflection in a line called the mirror line.

Reflection in a Line

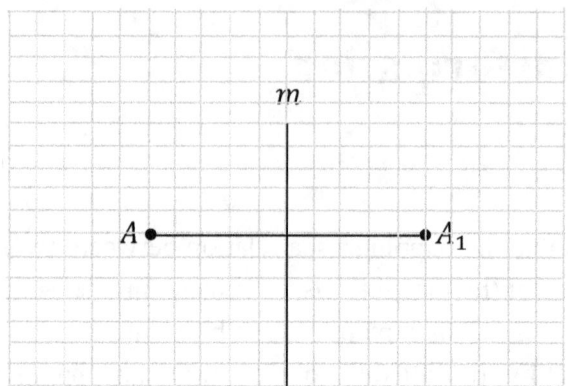

Figure 7.1

Figure 7.1 shows the reflection of a point A in a line m, called the mirror line. The point A and its image A_1 are the same distance from the mirror line. The line joining the point to its image is perpendicular to the mirror line.

Reflecting a Figure in a Line

To reflect an object, reflects each corner on the shape and then draw the reflection by joining the points you have reflected. Figure 7.2 shows the reflection of $\triangle XYZ$ in the lime m to give $X_1Y_1Z_1$. The points X, Y and Z and the corresponding images X_1, Y_1 and Z_1 are the same perpendicular distance from the mirror line.

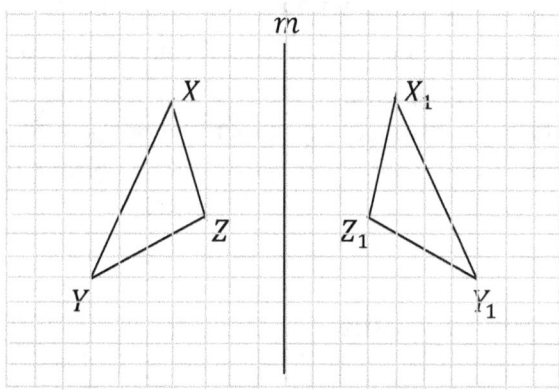

Figure 7.2

Notice that the image has the same shape and size as the original object. The object and image are said to be congruent.

Constructing a Reflected Image

The following steps can be used to reflect a figure in a line:

1. Draw a perpendicular line from each corner to the mirror line

2. Measure the distance of each corner from the mirror line and measure the same distance on the other side

3. Join the points you have reflected to from the image

Try this 1

On graph paper draw axes with x-axis from -5 to 5 and y-axis from -6 to 6, using a scale of 2 cm to 1 unit on both axes

(a) Plot the points A (3, 1), B (4, 2) and C (2, 3). Join the points A, B and C to form triangle ABC.

(b) (i) Reflect A in the x-axis. Label the image A_1 and write down its coordinates.

(ii) Reflect B and C in the x-axis. Label the images of B and C, B_1 and C_1 respectively.

(iii) Write down the coordinates of B_1 and C_1. Join A_1, B_1 and C_1 to form triangle $A_1B_1C_1$.

(c) (i) Reflect $\triangle ABC$ in the y-axis. Label the image A_2 B_2 C_2.

(ii) Write down the coordinates of A_2, B_2 and C_2.

(d) (i) Draw the line $y = x$

(ii) Reflect $\triangle ABC$ in the line $y = x$. Label the image $A_3 B_3 C_3$.

(iii) Write down the coordinates of A_3, B_3 and C_3

(e) (i) Draw the line $y = -x$

(ii) Reflect $\triangle A_2 B_2 C_2$ in the line $y = -x$. Label the image $A_4 B_4 C_4$

(iii) Write down the coordinates of A_4, B_4 and C_4.

(f) (i) Reflect $\triangle A_1 B_1 C_1$ in the line $x = 2$. Label the image $A_5 B_5 C_5$.

(ii) Write down the coordinates of A_5, B_5 and C_5.

(g) (i) Reflect $\triangle A_2 B_2 C_2$ in the line $y = -1$. Label the image $A_6 B_6 C_6$

(ii) Write down the coordinates of A_6, B_6 and C_6.

Reflection in the XY- Plane

In the coordinate plane, the mirror line may be given as an equation or it might just be the x- axis or the y - axis. The results of the exercises in Try this 1 are summarised below.

Transformation	Mapping
1. Reflection in the x- axis	$(x, y) \rightarrow (x, -y)$
2. Reflection in the y- axis	$(x, y) \rightarrow (-x, y)$
3. Reflection in the line $y = x$	$(x, y) \rightarrow (y, x)$
4. Reflection in the line $y = -x$	$(x, y) \rightarrow (-y, -x)$
5. Reflection in the line $x = a$	$(x, y) \rightarrow (2a - x, y)$

6. Reflection in the line $y = a$ $\qquad$ $(x, y) \rightarrow (x, 2a - y)$

Example

Find the image formed by reflecting (- 3, 2) in the line $y = x$

The image of the point (x, y) when reflected in the line $y = x$ is (y, x). So the image of (- 3, 2) is (2, - 3).

Try this 2

Find the image formed by reflecting (2, - 3) in each of the following mirror lines:

(a) the line $y = 0$ $\qquad$ (b) the line $x = 0$

(c) the line $y = x$ $\qquad$ (d) the line $y = -x$

Example

Find the image formed by reflecting (1, 3) in the line $x = -2$

The image of the point (x, y) after reflection in the line $x = a$ is $(2a - x, y)$.

So, the image of (1, 3) is $(-5, 3)$

Try this 3

Find the image formed by reflecting (- 2, 1) in each of the following mirror lines:

(a) the line $x = -3$ $\qquad$ (b) the line $x = 1$

(c) the line $y = 2$ $\qquad$ (d) the line $y = -1$

Exercise 7.1

1. Find the image of the following points after reflection in the x-axis

(a) (3, 5)   (b) (- 4, - 7)   (c) (5, - 2)   (d) (0, - 6)   (e) (- 3, 2)

2. Find the image of the following points after reflection in the y-axis

(a) (5, 7)   (b) (- 7, - 4)   (c) (2, - 5)   (d) (6, 0)   (e) (- 2, 3)

3. Find the image of the following points after reflection in the line $y = x$

(a) (2, 3)   (b) (- 3, 4)   (c) (5, - 2)   (d) (7, 0)   (e) (- 1, - 6)

4. Find the image of the following points after reflection in the line $y = -x$

(a) (5, 3)   (b) (- 2, 3)   (c) (3, - 4)   (d) (- 4, - 5)   (e) (0, - 2)

5. Find the image of the following points after reflection in the line $x = 2$

(a) (4, - 3)   (b) (- 2, - 1)   (c) (6, 5)

6. Find the image of the following points after reflection in the line $y = -3$

(a) (5, - 2)   (b) (3, 4)   (c) (- 8,-6)

7. State the mirror line for the following points and their reflection.
(a) (5, - 2); (5, 2)   (b) (4, 3); (3, 4)

(c) (- 7, 4); (7, 4)   (d) (- 7, - 2); (2, 7)

8. State the mirror line for the following points and their reflection

(a) (2, 3); (4, 3)   (b) (- 2, 1); (- 2, 3)   (c) (4, 0); (8, 0)

(d) (- 7, -2); (- 7, 4)   (e) (3, - 10); (3, 6)

7.2 Rotation

A rotation is a transformation that turns an object about a fixed point, called the centre of rotation. When describing a rotation, you will need to state the centre, the size of the angle and the direction of rotation.

A rotation can be either clockwise or anticlockwise. Clockwise rotations are negative and anticlockwise rotations are positive. A rotation of 180^0 is called $\frac{1}{2}$ turn, a rotation of 90^0 is called $\frac{1}{4}$ turn and a rotation of 270^0 is called $\frac{3}{4}$ turn.

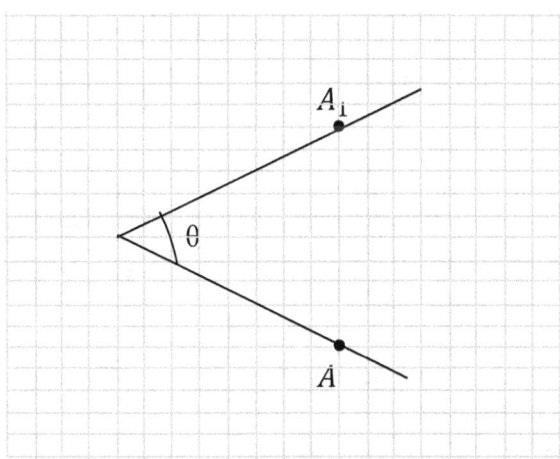

Figure 7.3

Figure 7.3 shows a point A rotated through angle θ anticlockwise to form the image A_1. The line OA has moved to OA_1. The angle of rotation is the angle between OA and OA_1. Note that:

1. The points A and A_1 are the same distance from the centre of rotation.

2. The centre of rotation lies on the perpendicular bisector of AA_1.

Rotating a set of points

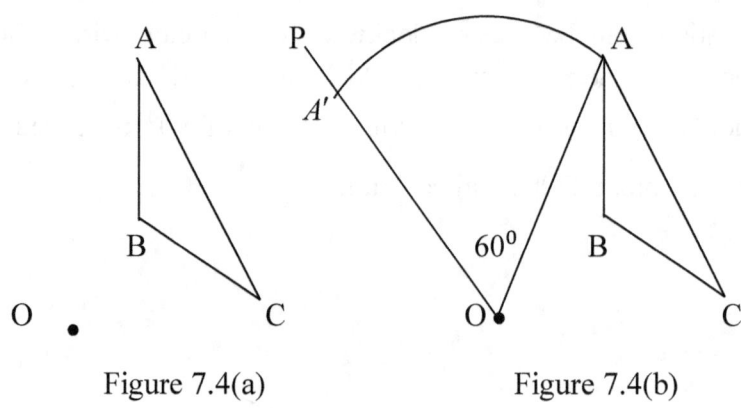

Figure 7.4(a)   Figure 7.4(b)

Figure 7.4(a) shows a triangle, ABC, and a point O, the centre of rotation. You can construct the image of this triangle about O through 60^0 by following the steps below:

1. Draw a line from O to A, and then draw angle $AOP = 60^0$ as shown in Figure 7.4(b).

2. With centre O, draw an arc from A to cross OP at A'. (See Figure 7.4(b))

The images of B and C are constructed by similar method. Join the image points to form triangle $A'B'C'$, as shown in Figure 7.5

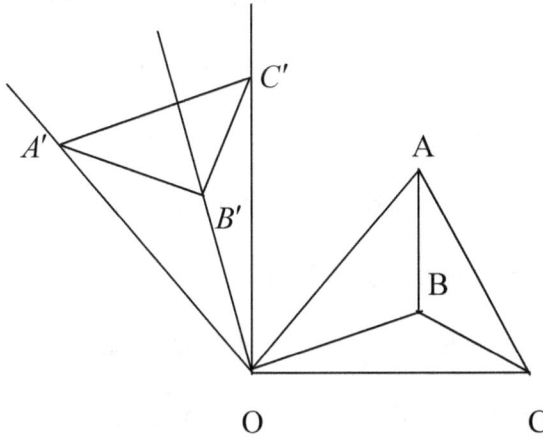

Figure 7.5

Note: The angle between the line joining any vertex and the line joining its image is 60^0

Generally, it is important to note the following:

1. A point and its image are the same distance from the centre of rotation.

2. The lines from the centre of rotation to any point turn through the same angle.

3. The image has the same size and shape as the object.

Try this 4

On graph paper draw axes with x-values from – 5 to 5 and y- axis from – 6 to 6. Use a scale of 2 cm to 1 unit on both axes.

(a). Plot the points A (2, 2), B (2, 3) and C (3, 2)

(b). (i) Rotate A(2, 2) by 90^0 anticlockwise about the origin. Label the image A' and write down its coordinates.

(ii) Rotate B and C by 90^0 anticlockwise about the origin. Label the corresponding images B' and C'. Join A', B' and C' to form triangle $A'B'C'$.

(c). (i) Rotate triangle ABC by 180^0 anticlockwise about the origin. Label the image $A''B''C''$.

(ii) Write down the coordinates of A'', B'' and C''.

(d). (i) Rotate triangle ABC by 270^0 anticlockwise about the origin. Label the image $A'''B'''C'''$

(ii) Write down the coordinates of A''', B''' and C'''

The results of the above exercises are summarised below.

Transformation	**Mapping**
1. Rotation through 90^0 about the origin	$(x,y) \rightarrow (-y,x)$
2. Rotation through 180^0 about the origin	$(x,y) \rightarrow (-x,-y)$
3. Rotation through 270^0 about the origin	$(x,y) \rightarrow (y,-x)$

Note that a 270^0 rotation in an anticlockwise direction is the same as a 90^0 rotation in a clockwise direction.

Example

Find the image formed by rotating (3, - 2) 90^0 anticlockwise about the origin.

The image of the point (3, - 2) is (2, 3)

Try this 5

Find the image formed by rotating (- 3, 1) anticlockwise through each of the following angles about the origin.

(a) 90^0 (b) 180^0 (c) 270^0

Exercise 7.2

1. Under a rotation of 90^0 about the origin, find the image of each of

the following points

(a) (3, 5)   (b) (- 1, 0)   (c) (- 2, -3)   (d) (0, 4)   (e) (5, - 3)

2. Under a rotation of 180^0 about the origin, find the image of each

 of the following points.

(a) (- 2, 3)   (b) (4, - 2)   (c) (- 3, - 1)   (d) (2, 1) (e) (0, - 4)

3. Under a rotation of 270^0 about the origin, find the image of each of

the following points.

(a) (3, 2)   (b) (0, - 3)   (c) (- 5, - 4)   (d) (- 6, 3)   (e) (3, 0)

7.3 Translation

A translation moves an object from one place to another without changing its size and shape.

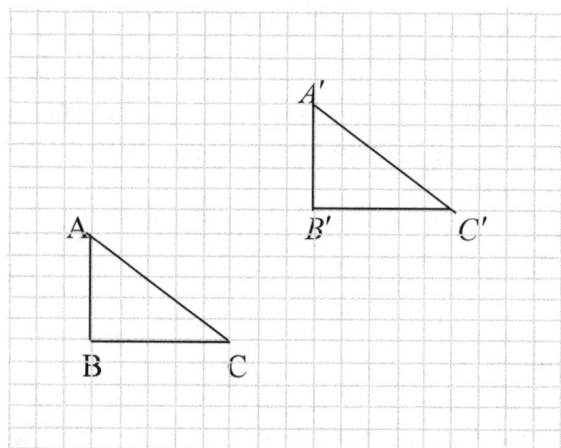

Figure 7.6

The triangle ABC has been translated to give the image $A'B'C'$. Observe that every vertex of triangle ABC moves 8 units to the right and 6 units up. The horizontal displacement followed by the vertical displacement is written as the vector $\begin{pmatrix} 8 \\ 6 \end{pmatrix}$ called the translation (or the displacement) vector. The horizontal displacement is written at the top of the vector and the vertical displacement at the bottom. A move downwards or to the left is indicated by a negative sign and a move up or to the right is indicated by positive sign.

Note that:

1. Every point of the object is moved the same distance and in the same direction

2. The image has the same size and shape as the object

Try this 6

On graph paper draw axes with x-values from – 5 to 5 and y-values from – 6 to 6. Use a scale of 2 cm to represent 1 unit on both axes.

(a). Plot A (2, 1), B (3, 1) and C (3, 2). Join A, B and C to form triangle ABC.

(b). (i) Translate the point A by vector $\begin{pmatrix} 1 \\ 2 \end{pmatrix}$. Label the image A' and write down its coordinates

(ii) Translate the points B and C by $\begin{pmatrix} 1 \\ 2 \end{pmatrix}$ Label the corresponding images B' and C'. Write down the coordinates of B' and C'

(iii) Join A', B' and C' to form a triangle

(c). Translate triangle ABC by $\begin{pmatrix} -2 \\ -1 \end{pmatrix}$. Label the image $A''B''C''$

(d). (i) Plot $A'''(-3,2)$, $B'''(-2,2)$ and $C'''(-2,3)$

(ii) Find the translation vector that will translate triangle $A''B''C''$ onto triangle $A'''B'''C'''$

The result of the above exercises is summarised below.

Transformation	Mapping
Translation by the vector $\begin{pmatrix} a \\ b \end{pmatrix}$	$(x,y) \rightarrow (x + a, y + b)$

Example

Find the image formed when (4, - 3) is translated by the vector $\begin{pmatrix} -5 \\ 4 \end{pmatrix}$

The image of (4, - 3) is (- 1, 1)

Try this 7

Find the images of (5, 3) after translation by each of the following vectors

(a) $\begin{pmatrix} -2 \\ 1 \end{pmatrix}$ (b) $\begin{pmatrix} -3 \\ -5 \end{pmatrix}$ (c) $\begin{pmatrix} 0 \\ -4 \end{pmatrix}$

Exercise 7.3

1. Find the images of the following points

(a) (1, 3)   (b) (- 2, 5)   (c) (2, 0)   (d) (- 4, - 3)

after translation by each of the following vectors

(i) $\begin{pmatrix} 3 \\ 2 \end{pmatrix}$ (ii) $\begin{pmatrix} -2 \\ -1 \end{pmatrix}$ (iii) $\begin{pmatrix} 2 \\ -3 \end{pmatrix}$ (iv) $\begin{pmatrix} 0 \\ -1 \end{pmatrix}$

2. Find the points whose images are

(a) (5, 3)   (b) (- 4, 2)   (c) (- 6, - 5)   (d) (0, - 3)

after translation by each of the following vectors

(i) $\begin{pmatrix} 2 \\ 1 \end{pmatrix}$ (ii) $\begin{pmatrix} -2 \\ 3 \end{pmatrix}$ (iii) $\begin{pmatrix} -3 \\ -2 \end{pmatrix}$ (iv) $\begin{pmatrix} -2 \\ 0 \end{pmatrix}$

3. Find the translation vector of the following points and their images

(a) (5, 3); (3, 5)   (b) (3, - 10); (5, - 6)   (c) (9, -7); (4, - 9)

(d) (- 7, - 2); (2, - 6)   (e) (- 1, 5); (- 5, 2)

7.4 Enlargement

All the transformations you have studied so far in this chapter produce images that had the same size and shape as the object. An enlargement produces an image which is similar to the object. The image can be either larger or smaller than the original object but all angles remain the same. When you describe an enlargement, you will need to state the scale factor and the centre of enlargement.

Reduction

An enlargement that produces a reduced image is called reduction. The value of the scale factor determines whether an enlargement is a reduction. If the scale factor is between -1 and 1 the image will be smaller than the object.

Enlarging a Plane Figure

Figure 7.7 shows triangle ABC and the point O, the centre of enlargement.

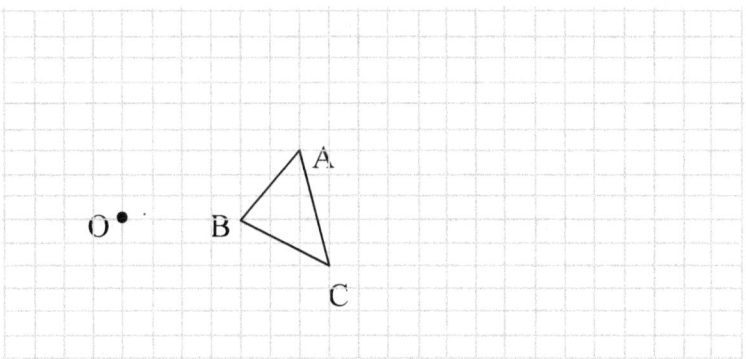

Figure 7.7

Triangle ABC has been enlarged by scale factor 2 to form the image $A_1 B_1 C_1$ as shown in Figure 7.8

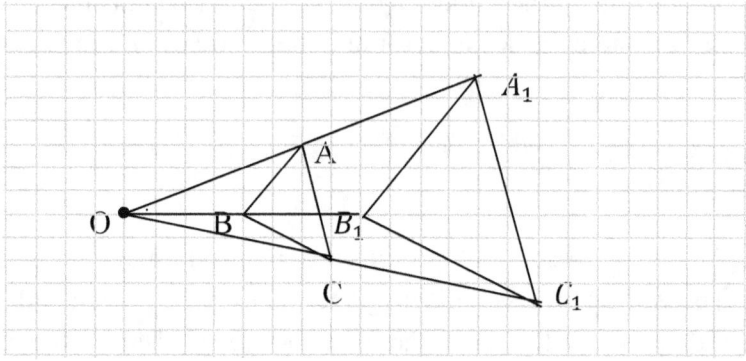

Figure 7.8

Recall that an enlargement produces a similar figure.

So $\dfrac{A_1B_1}{AB} = \dfrac{A_1C_1}{AC} = \dfrac{B_1C_1}{BC}$

Each of these ratios gives the scale factor of the enlargement. The scale factor can also be found by the following ratios:

$\dfrac{OA_1}{OA}$, $\dfrac{OB_1}{OB}$ and $\dfrac{OC_1}{OC}$

The scale factor determines the size of the image as compared to the size of the original figure.

Negative Enlargement

An enlargement that has the object and its image on opposite sides of the centre of enlargement is called negative enlargement. The scale factor is calculated as above but written with a negative sign before the number. In Figure 7.9 triangle ABC has been enlarged by scale factor -1 to give triangle $A_1B_1C_1$.

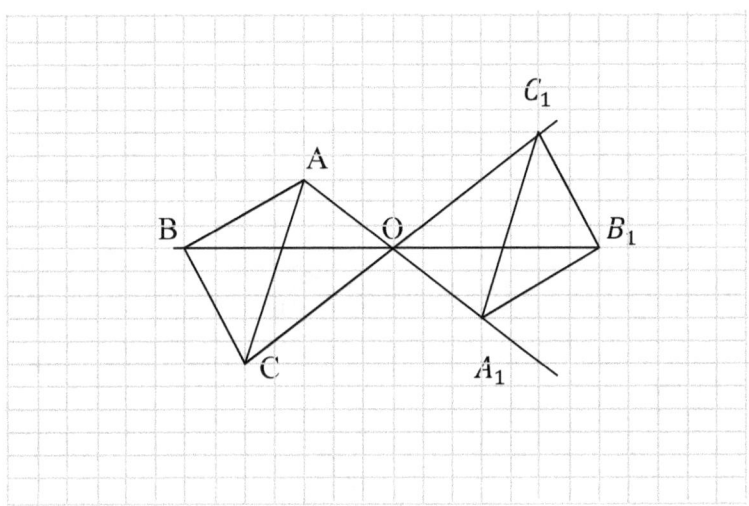

Figure 7.9

Try this 8

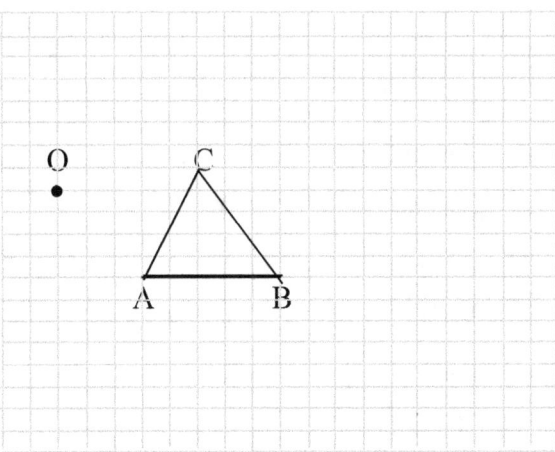

On graph paper, copy the diagram. Then enlarge the triangle ABC by scale factor 2 from the centre O. Label the image formed $A'B'C'$

(a) Measure the lines $A'B'$, $B'C'$ and $A'C'$

(b) Find the ratios $\dfrac{A'B'}{AB}$, $\dfrac{B'C'}{BC}$ and $\dfrac{A'C'}{AC}$

Example

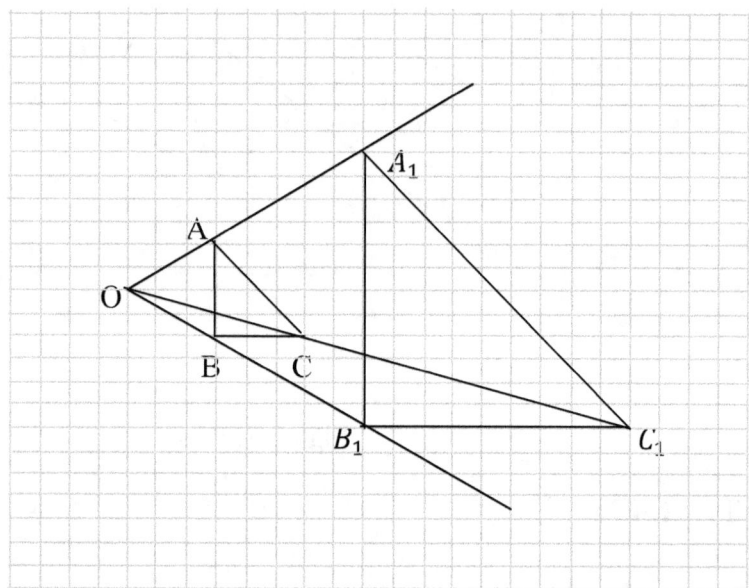

Triangle ABC has been enlarged from O to give triangle $A_1B_1C_1$. Find the scale factor and the length of A_1C_1

The scale factor is the ratio of the length of a side on the image to the length of the corresponding side on the object.

Scale factor $= \dfrac{A_1B_1}{AB} = \dfrac{12}{4} = 3$

The length of AC is 5

So the length of $A_1C_1 = 3 \times 5 = 15$

Try this 9

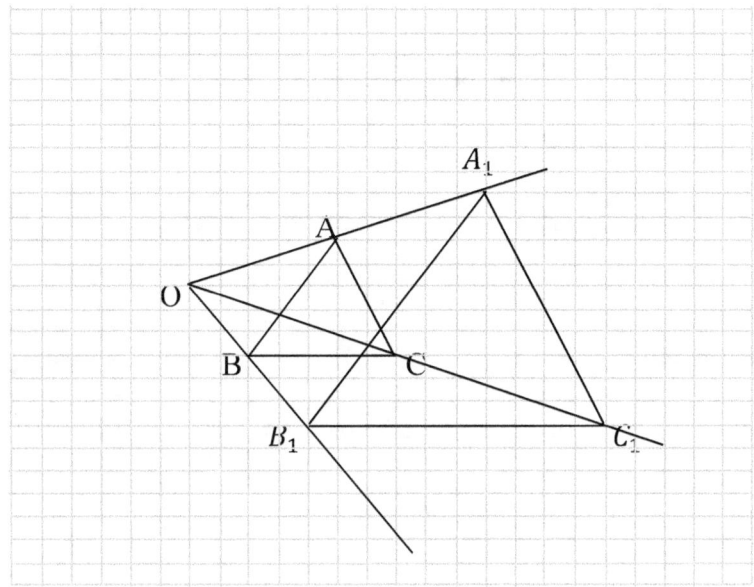

In the diagram above, triangle ABC has been enlarged to give triangle $A_1B_1C_1$. Find the scale factor.

Try this 10

On graph paper, draw axes with x-values from -4 to 6 and y-values from -6 to 6. Use a scale of 2 cm to represent 1 unit on both axes.

(a) Plot A (1, 2), B (3, 3) and C (3, 1). Join A, B and C to form a triangle

(b) . Enlarge triangle ABC by scale factor of $1\frac{1}{2}$ about the origin. Label the image $A'B'C'$

(c) Enlarge triangle ABC by scale factor -1 about (3, -1). Label the image $A''B''C''$

(d) Draw triangle $A'''B'''C'''$ with vertices at $A'''(1,-4)$, $B'''(-1,-5)$

and $C'''(-1,-3)$

(e) Triangle $A'''B'''C'''$ is an enlargement of triangle ABC. Find

the centre of enlargement.

The result of the exercise in Try 10 is summarised below.

Transformation	**Mapping**
Enlargement from the origin with scale factor k	$(x,y) \rightarrow (kx, ky)$

Example

Triangle ABC has vertices at A (3, 2), B (1, 3) and C (2, 4). It is enlarged to triangle $A'B'C'$ by scale factor 2 about the origin. Write down the vertices of triangle $A'B'C'$.

The vertex A' is (6, 4)

The vertex B' is (2, 6)

The vertex C' is (4, 8)

Try this 11

Triangle ABC has vertices at A (- 4, 2), B (2, - 2) and C (4, 6). It is enlarged to triangle $A'B'C'$ by scale factor of $-1\frac{1}{2}$ about the origin. Write down the vertices of triangle $A'B'C'$

Exercise 7.4

1. Draw a rectangle ABCD with sides AB = 3 cm and BC = 4 cm. Enlarge ABCD by a scale factor of – 1 about a point outside it. What is another way of describing this transformation?

2. Draw any triangle, and enlarge it about a point outside it with scale factors:

(a) $\frac{1}{2}$ (b) – 2   (c) $1\frac{1}{2}$

3.

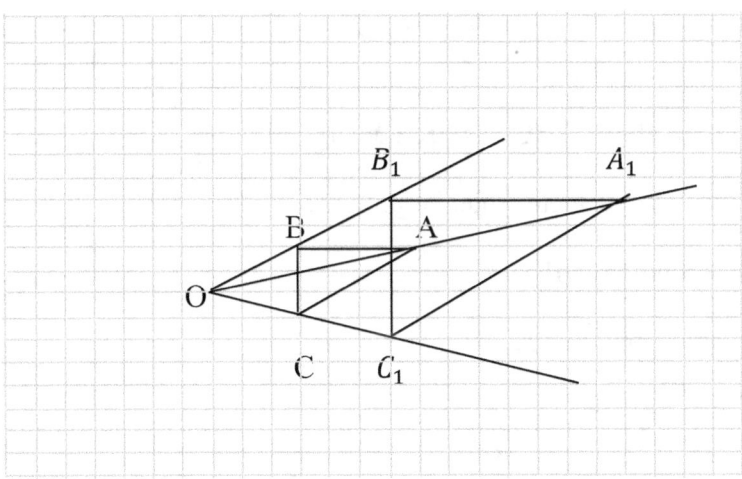

The triangle ABC is enlarged from O to $A_1B_1C_1$

(a) What is the scale factor of enlargement?

(b) Find the length of AC

(c) What is the length of A_1B_1?

4.

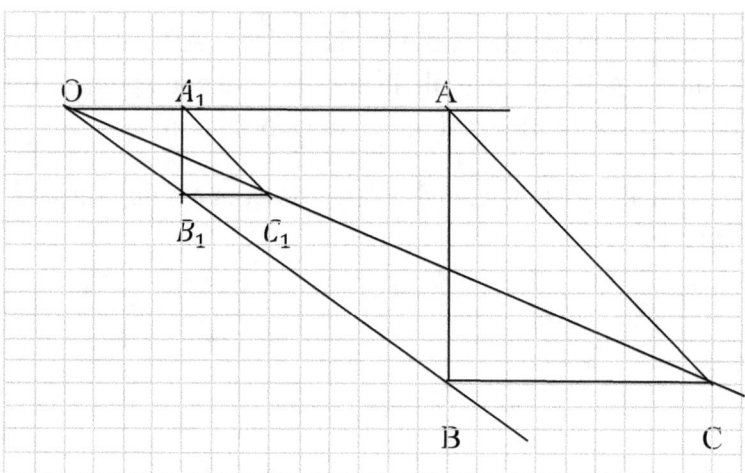

In the diagram above ABC has been enlarged to give $A_1B_1C_1$.

(a) What is the scale factor of enlargement?

(b) Find the lengths A_1B_1

(c) What is the length of AC?

5.

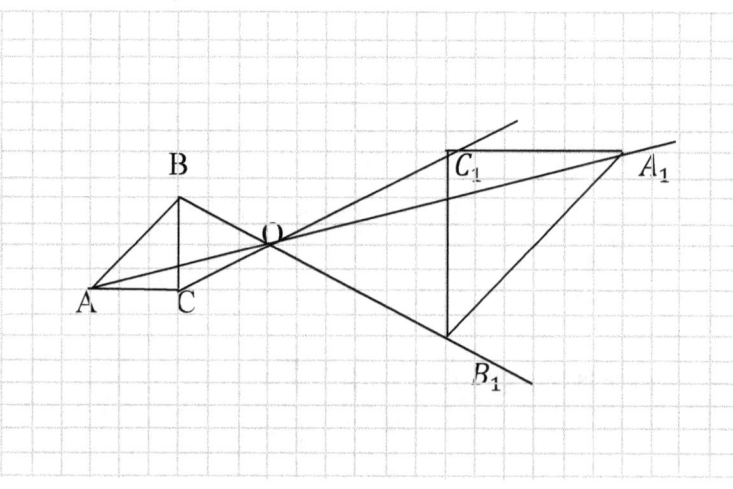

In the diagram triangle ABC has been enlarged to give $A_1B_1C_1$.

(a) Find the scale factor

(b) Find the length of A_1B_1

6. Write down the images of the following points after enlargement from the origin with the given scale factor

(a) (2, 3); 2   (b) (- 4, 6); $\frac{1}{2}$ (c) (3, - 4); - 1   (d) (0, - 6); $-\frac{1}{3}$

(e) (- 1, - 2); 3

7. Write down the image of each of the following points after enlargement from the centre (5, 4) with scale factor 2

(a) (3, 3)   (b) (- 6, 2)   (c) (5, 6)   (d) (- 3, - 2)

8. Write down the image of each of the following points after an enlargement from the centre (1, 1) with scale factor – 3

(a) (1, 2)   (b) (4, - 4)   (c) (- 2, 1)   (d) (- 3, -2)

9. A triangle ABC has vertices at A (3, - 2), B (- 1, 2) and C (- 3, - 1). It is enlarged to $A'B'C'$ with vertices at $A'(-6,4)$, $B'(2,-4)$ and $C'(6,2)$.

(a) Draw the triangles on graph paper

(b) Find the centre of enlargement.

(c) Find the scale factor

10. A triangle has vertices at A (1, 1), B (3, 2) and C (3, 1). It is enlarged to $A'B'C'$ with vertices at $A'(-1,-1)$, $B'(5,2)$ and $C'(5,-1)$.

(a) Draw the triangles on graph paper

(b) Find the centre of enlargement.

(c) Find the scale factor

7.5 Similar Figures

Two plane figures are similar if they have all corresponding sides in the same proportion. This means they have the same shape, but may be different size. Similar figures have corresponding angles equal. Recall that when a figure is enlarged the new figure is similar to the original figure.

Ratio of Areas

Consider the two similar rectangles shown in Figure 7.10.

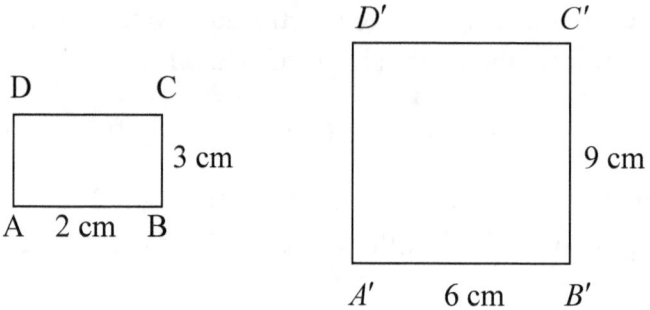

Figure 7.10

By definition each pair of corresponding sides are in the same ratio. Clearly

$$\frac{A'B'}{AB} = \frac{B'C'}{BC} = 3$$

The lengths of the two rectangles are in the ratio 3 : 1. Notice that each side of ABCD has been multiplied by 3 to obtain the sides of $A'B'C'D'$.

$$\frac{Area\ of\ A'B'C'D'}{Area\ of\ ABCD} = \frac{6 \times 9}{2 \times 3} = 9$$

The ratio of the areas is 9 : 1, which is the square of the ratio of the sides. Therefore the area of $A'B'C'D'$ is 9 times the area of ABCD.

This example illustrates that the ratio of the area of similar figures is the square of the ratio of the sides. For example, if the sides are in the ratio $k:1$ then the areas will be in the ratio $k^2:1$.

Example

Two rectangles ABCD and $A'B'C'D'$ are similar. The sides of $A'B'C'D'$ are twice the sides of ABCD. If the area of ABCD is 20 cm^2, find the area of $A'B'C'D'$

The sides of $A'B'C'D'$ are 2 times the sides of ABCD, hence the area of $A'B'C'D'$ is 4 times the area of ABCD i.e. $4 \times 20 = 80$ cm^2

Try this 12

The radii of two spheres are 2 cm and 6 cm. If the area of the bigger sphere is 36 cm^2, find the area of the smaller sphere.

Exercise 7.5(a)

1. The area of two similar quadrilaterals are 25 cm^2 and 225 cm^2. What is the ratio of their lengths?

2. Two spheres have areas 88 cm^2 and 264 cm^2. If the radius of the smaller sphere is 7 cm, find the radius of the bigger sphere.

3. The area of two similar triangles ABC and $A'B'C'$ are 28 cm^2 and 112 cm^2 respectively. If the height of triangle $A'B'C'$ is 10 cm, find the height of triangle ABC.

4. The area of two rectangles ABCD and $A'B'C'D'$ are 20 cm^2 and 180 cm^2 respectively. If the length $B'C'$ is 15 cm, find the length of AB.

5. A triangle is enlarged in the ratio 1: 3. If the area of the triangle is 12 cm^2, find the area of the enlarged triangle.

6. A photograph is enlarged in the ratio 2: 3. If a building occupies 10 cm^2 of the original photo, how much area does it occupy in the enlargement?

7. The area of a rectangular plot of land is 48 cm^2. If the plot is divided into smaller rectangular plots which lengths are $\frac{1}{4}$ of that of the bigger plot, find the area of a smaller plot.

8. The plan for a building has a scale of 2 cm to 5 cm. If the area of the floor of a room is 50 m^2, find the area of the floor on the plan.

9. The cost of painting a rectangular wall 8 m by 5 m is GH¢ 25 per m^2. How much does it cost to paint a similar wall whose lengths are 4 times longer?

10. The length of a rectangular floor is 1.5 m. How many rectangular tiles, each of length 5 cm, would be needed to tile the floor?

Ratio of Volumes

Consider the two similar cuboids shown in Figure 7.11.

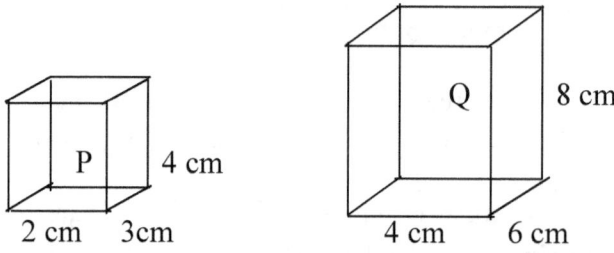

Figure 7.11

Notice that each side of P is multiplied by 2 to obtain the sides of Q.

$$\frac{Volume\ of\ Q}{Volume\ of\ P} = \frac{4 \times 6 \times 8}{2 \times 3 \times 4} = 8$$

The ratio of the volumes is 8 : 1, which is the cube of the ratio of the lengths of the sides. The volume of Q is 8 times the volume of P.

This example illustrates that the ratio of the volume of similar solids is the cube of the ratio of the sides. For example, if the sides are in the ratio $k: 1$ then the volumes will be in the ratio $k^3: 1$.

Example

The volume of a cuboid P is 6 cm³. Find the volume of a similar cuboid which sides are twice the sides of P.

The sides of the bigger cuboid are 2 times the sides of P. Hence its volume is 8 times the volume of P.

So the volume of the bigger cuboid = $8 \times 6 = 48$ cm³

Try this 13

The volume of two similar cylindrical containers P and Q are 35.2 cm³ and 281.6 cm³ respectively. If the height of P is 1.4 cm, find the height of Q.

Exercise 7.5(b)

1. The volume of two similar cylinders, P and Q are 4 cm³ and 108 cm³ respectively. Find the ratio of the radius, if Q is an enlargement of P.

2. A container P has a rectangular base 3 cm by 2 cm and height 1 cm. Find the volume of a similar container Q which height is 3 cm.

3. The volume of two cylindrical containers P and Q are 19.8 cm^3 and 158.4 cm^3 respectively. If the radius of P is 3 cm, find the radius of Q.

4. The base radius of two circular cylinder P and Q are 3 cm and 9 cm respectively. If the volume of P is 198 cm^3, find the volume of Q.

5. A water container of height 6 cm has a volume of 25 cm^3. What is the height of a similar container of volume 200 cm^3.

6. The volume of two similar cones P and Q are 77 cm^3 and 616 cm^3 respectively. If the height of P is 1.5 cm and base radius of Q is 14 cm find the slant height of Q.

7. A rectangular tank with a horizontal base and height 2 m, contains 24 m^3 of water. If a similar rectangular tank contains 192 m^3 of water, find the depth of water.

8. The height of a cylindrical container is 9 cm. A tap fills the container at a rate of 8 cm^3 s^{-1}. If the depth of water in the container after 3 s is 6 cm, find the volume of the container.

9. A circular cylinder which leaks has an internal diameter 12 cm, height 10 cm and it was filled with water. If a similar circular cylinder of internal diameter 6 cm which catches water from the container is filled in 3 minutes, find the rate of leakage in cm^3 s^{-1}.

10. The base of a tank is a square, side 6 cm and the tank is 4 cm high. If water is poured from the tank into similar tanks which base is a square of side 3 cm, how many of such tanks can be filled?

Review exercise 7

1. Find the image of each of the following points

(i) (2, 4)   (ii) (- 3, - 5)   (iii) (3, - 2)

under reflection in the

(a) y axis   (b) x axis   (c) line $y = x$ (d) line $y = -x$

2. Find the image of each of the following points

(i) (- 3, 2)   (ii) (4, - 5)   (iii) (- 1, - 2)

under reflection in the line

(a) $x = 3$ (b) $x = -2$ (c) $y = -3$ (d) $y = 2$

3. State the mirror line for each of the following reflection:

(a) $(3,2) \rightarrow (-3,2)$ (b) $(1,-3) \rightarrow (1,3)$

(c) $(4,3) \rightarrow (3,4)$ (d) $(-3,-2) \rightarrow (2,3)$

4. State the mirror line for each of the following reflection:

(a) $(3,2) \rightarrow (1,2)$ (b) $(3,4) \rightarrow (3,2)$

(c) $(-1,-2) \rightarrow (-3,-2)$ (d) $(-3,2) \rightarrow (-3,-4)$

5. Find the image of each of the following point

(i) (- 3, 0)   (ii) (4, - 3)   (iii) (- 5, 4)

when rotated about the origin through

(a) 90^0 (b) 180^0 (c) 270^0

6. Find the image of each of the following points

(i) (0, 1)   (ii) (- 4, 3)   (iii) (5, 3)

when rotated about the origin through

(a) - 90⁰   (b) - 180⁰   (c) - 270⁰

7. Find the image of each of the following points

(i) (2, -1)   (ii) (- 2, 4)   (iii) (- 3, - 4)

under the translation by the vector

(a) $\binom{3}{2}$ (b) $\binom{-4}{3}$ (c) $\binom{-2}{-3}$ (d) $\binom{5}{-2}$

8. Find the translation vectors for each of the following translations

(a) $(3,2) \rightarrow (4,-1)$ (b) $(-2,1) \rightarrow (-3,3)$

(c) $(2,3) \rightarrow (-1,2)$ (d) $(0,2) \rightarrow (-3,5)$

9. Find the image of each of the following points

(i) (2, 4)   (ii) (- 4, 6)   (iii) (- 6, - 2)

after enlargement from the origin with scale factor

(a) 2   (b) $\frac{1}{2}$ (c) $-\frac{3}{2}$

10. (a) Using a scale of 2 cm to 2 units on each axis, draw on a graph

     paper two perpendicular axes Ox and Oy, for the intervals

     $-10 \leq x \leq 10$ and $-10 \leq y \leq 10$

   (b) Draw triangle ABC with vertices A (3, 2), B (6, 5) and

     C (2, 7)

(c) Draw the image triangle $A'B'C'$ of triangle ABC under an anticlockwise rotation of 90^0 about the origin

(d) Draw the image triangle $A''B''C''$ of triangle $A'B'C'$ under a translation by the vector $\begin{pmatrix} 0 \\ -8 \end{pmatrix}$

(e) Draw the image triangle $A'''B'''C'''$ of triangle $A''B''C''$ under an enlargement from (0, -4) with scale factor – 1

(f) Describe the single transformation which maps triangle $A'''B'''C'''$ onto triangle $A'B'C'$

11. Draw on graph paper two perpendicular axes Ox and Oy for $-10 \le x \le 10$ and $-10 \le y \le 10$, using a scale of 2 cm to 2 units on both axes

(a) Plot the points A (2, 8), B (2, 3) and C (6, 3). Describe triangle ABC

(b) Draw triangle $A'B'C'$, which is the image of triangle ABC under a clockwise rotation through 90^0 about the origin (0, 0),

(c) Draw triangle $A''B''C''$ which is the reflection of triangle $A'B'C'$ in the line $y - x = 0$

(d) What single transformation maps triangle $A''B''C''$ onto triangle ABC.

12(a) Draw on graph paper two perpendicular axes Ox and Oy for the intervals $-10 \leq x \leq 10$ and $-10 \leq y \leq 10$, using a scale of 1 cm to 1unit on both axes.

(b) Given A (2, 5) and the vectors $\overrightarrow{AB} = \begin{pmatrix} 2 \\ 2 \end{pmatrix}$ and $\overrightarrow{BC} = \begin{pmatrix} 2 \\ -3 \end{pmatrix}$ draw triangle ABC, showing clearly the coordinates of all vertices

(c) Draw triangle $A'B'C'$ with vertices $A'(-1,2)$, $B'(1,4)$ and $C'(3,1)$

(d) Triangle ABC is translated by a vector T to form triangle $A'B'C'$. Find the vector T.

(e) Draw the image triangle $A''B''C''$ of triangle ABC under a reflection in the x −axis

(f) Draw the image triangle $A'''B'''C'''$ of triangle $A''B''C''$ under a translation by the vector T.

(g) What single transformation maps triangle $A'''B'''C'''$ onto triangle $A'B'C'$

13. The base radius of two similar cones P and Q are 2.1 cm and 6.3 cm respectively. If the surface area of P is 66 cm^2 find the surface area of Q.

14. The surface areas of two similar cylindrical containers P and Q are 132 cm^2 and 528 cm^2 respectively. If the radius of Q is 2.8 cm, find the radius of P.

15. The heights of two similar cones P and Q are 2.8 cm and 5.6 cm respectively. If the volume of P is 26.4 cm^3, find the volume of Q.

16. The volume of two spheres P and Q are 26.4 cm^3 and 712.8 cm^3 respectively. If the radius of Q is 6 cm, find the radius of P.

17. An architect's model of a house is in the scale 1: 10. If the living - room is 800 m^3, what is the volume of the model of the living-room?

18. A photograph is enlarged in the ratio 2: 5. If the sky occupies 10 cm^2 of the original photo, how much area does it occupy in the enlargement?

Chapter Test 7

Take this test as you would take a test in a class. After you are done, check your work against the answers in the back of the book.

1. The triangle with vertices at A (2, 1), B (5, 4) and C (1, 4) is reflected in the x axis to form the image triangle $A'B'C'$. Then triangle $A'B'C'$ is reflected in the y axis to form triangle $A''B''C''$.

(a) Draw on graph paper triangle ABC, $A'B'C'$ and $A''B''C''$.

(b) What single transformation would map triangle $A''B''C''$ onto

triangle ABC

2. Triangle ABC, with vertices A(2, 2), B(2, 3) and C(0, 2) is

enlarged to form triangle $A'B'C'$, with vertices $A'(-1,-1)$,

$B'(-1,-3)$ and $C'(3,-1)$

(a) Draw on a graph paper triangle ABC and $A'B'C'$

(b) Find

(i) the centre of enlargement

(ii) the scale factor of enlargement

3. (a) Using a scale of 2 cm to 2 units on each axis, draw on graph

paper two perpendicular axes Ox and Oy, for the intervals

$-10 \le x \le 10$ and $-10 \le y \le 10$

(b) Draw triangle ABC with vertices at A (6, 8), B (8, 4) and

C (4, 4)

(c) Draw the image triangle $A'B'C'$ of triangle ABC under a

reflection in the y- axis

(d) Draw the image triangle $A''B''C''$ of triangle $A'B'C'$ under an

anticlockwise rotation of 180^0 about the origin

(e) Describe the single transformation which maps triangle

$A''B''C''$ onto triangle ABC

4. Using a scale of 1 cm to 1 unit on each axis, draw on graph paper two perpendicular axes Ox and Oy, for the intervals $-10 \leq x \leq 10$ and $-10 \leq y \leq 10$

(a) (i) Draw triangle ABC with vertices A(1,1), B(4,1) and C(4,3)

(ii) Draw the image triangle $A_1 B_1 C_1$ under a reflection of triangle ABC in the x- axis.

(iii) Draw the image triangle $A_2 B_2 C_2$ under a reflection of triangle $A_1 B_1 C_1$ in the line $y = -4$

(iv) Describe fully the single transformation which maps triangle ABC onto triangle $A_2 B_2 C_2$

(b) (i) Draw the image triangle $A_3 B_3 C_3$ of triangle ABC under a rotation about the origin through 180^0

(ii) Describe fully the single transformation which maps $A_3 B_3 C_3$ onto triangle $A_1 B_1 C_1$.

(c) Draw the image triangle $A_4 B_4 C_4$ of triangle ABC under enlargement from the centre P(8,1) with scale factor 2

5. X and Y lie on the sides AB and AC of triangle ABC. AX = 3 cm, XB = 4 cm AY = 6 cm, YC = 8 cm. If triangle AXY has area 4.5 cm^2, what is the area of XYCB.

6. Two similar cups are 24 cm and 4 cm high respectively. If the smaller cup holds $\frac{1}{8}$ litre of water, how much water would the bigger cup hold?

7. A model house is built with a scale 3 cm : 4 m. If the area of the floor of the living- room of the model is 0.45 m², what is the area of the living- room

8

Lengths and Areas of Plane Figures

Perimeter and Area

The perimeter of a plane figure is the distance around it. The perimeter of a polygon is the sum of the lengths of its sides.

The area of a plane figure is the number of square units it takes to fill the interior of that figure. The area of polygons can be found by use of appropriate formulas.

8.1 Triangles

The Perimeter of a Triangle

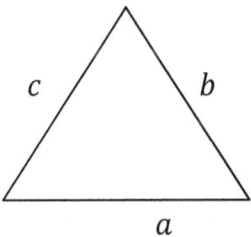

Figure 8.1

The perimeter of the triangle shown in Figure 8.1 is given by the sum $a + b + c$.

i.e. the $perimeter = a + b + c$

Example

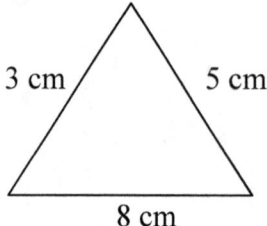

Find the perimeter of the triangle shown above

Perimeter $= 3 + 5 + 8$

$\qquad = 16$

The perimeter is 16 cm

Try this 1

The sides of a triangle are x, $2x$ and 13. If the perimeter of the triangle is 31 find the value of x.

Area of a Triangle

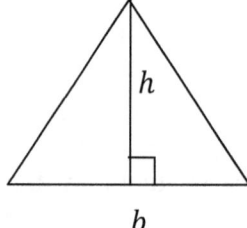 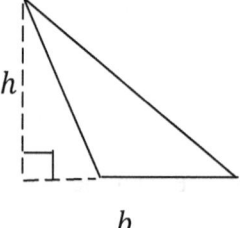

Figure 8.2

The area, A, of a triangle is given by the formula

$$A = \frac{1}{2}bh$$

where b is the length of the base and h is the height, the perpendicular distance from the base to the vertex.

Example

Calculate the area of the triangle shown in the diagram bellow

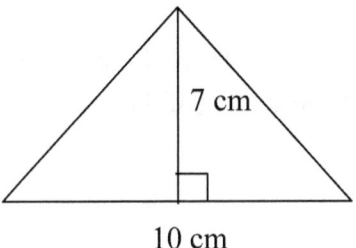

7 cm

10 cm

$$A = \frac{1}{2}bh$$

$$= \frac{1}{2} \times 10 \times 7$$

$$= 35$$

The area is 35 cm^2

Try this 2

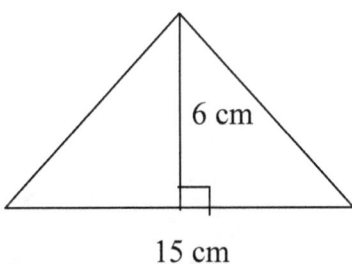

6 cm

15 cm

Calculate the area of the triangle shown in the diagram above

Exercise 8.1(a)

1. The perimeter of an equilateral triangle is 33 cm. Find the length of the sides?

2. The base of an isosceles triangle is 15 cm. If the perimeter of the triangle is 39 cm, find the length of the equal sides.

3. The height of an isosceles triangle is 4 cm. If lengths of the equal sides are each 5 cm find the perimeter of the triangle.

4.

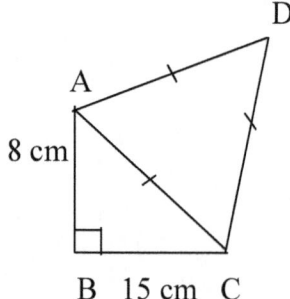

In the diagram above ABC is a right angled triangle and ACD is an equilateral triangle. If AB = 8 cm and BC = 15 cm, find the perimeter of ABCD

In Exercises 5 – 10, calculate the area of each triangle

5.

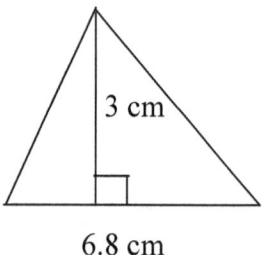

6.

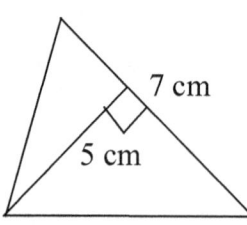

7.

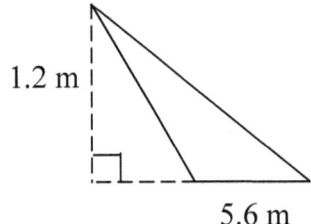

1.2 m

5.6 m

8.

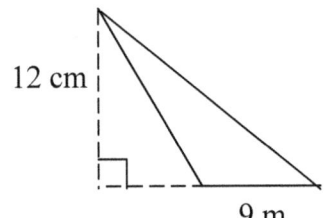

12 cm

9 m

9.

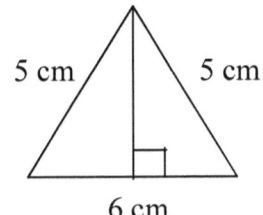

5 cm   5 cm

6 cm

10.

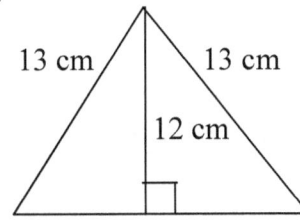

13 cm   13 cm

12 cm

11.

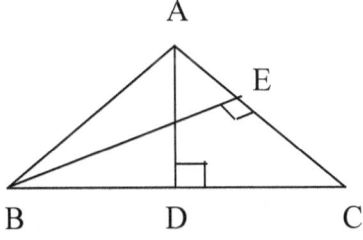

A

E

B   D   C

In the diagram shown above, BC = 5 m, AC = 4 m and AD = 8 m. Calculate the area of triangle ABC and the length of BE.

12. Triangle ABC is an equilateral triangle and BC = 10 cm. If N is the midpoint of BC, calculate AN and the area of the triangle

13. In triangle ABC $\angle ABC = 90°$, AB = 6 cm, BC = 8 cm and AC = 10 cm. Find the length of the altitude from vertex B to $\overline{AC}$.

14.

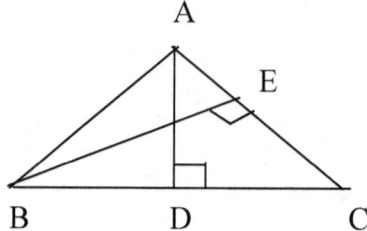

In the diagram shown above, triangle ABC is an isosceles triangle, AB = AC = 15 cm, AD = 9 cm. Calculate the area of triangle ABC and the length of BE.

The Trigonometric Formula

This formula is used when we are given the length of two sides of the triangle and the angle between these sides.

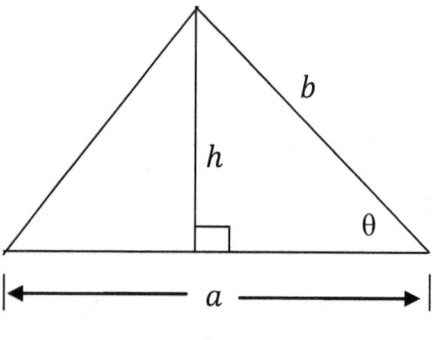

Figure 8.3

You can see from Figure 8.3 that

$$\frac{h}{b} = \sin \theta°$$

so $h = b \sin \theta°$

Hence the area of the triangle $= \frac{1}{2} ab \, sin \, \theta°$

Example

Calculate the area of the triangle shown below

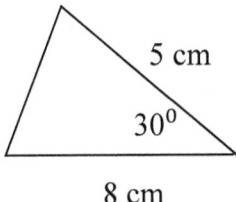

5 cm

$30°$

8 cm

$A = \frac{1}{2} ab \, sin\theta°$

$= \frac{1}{2} \times 8 \times 5 \times sin \, 30°$

$= 10$

The area is 10 cm $^{-2}$

Try this 3

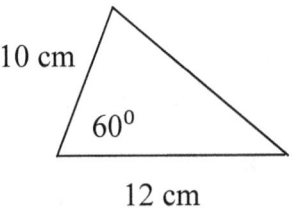

10 cm

$60°$

12 cm

Calculate the area of the triangle shown above.

Exercise 8.1(b)

In Exercises 1- 6, calculate the area of each of the following triangles

1.

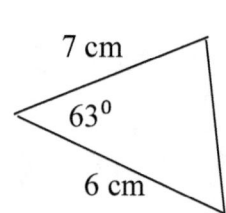

2.

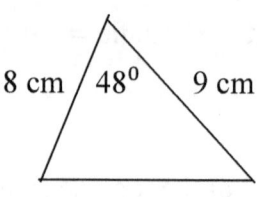

3.

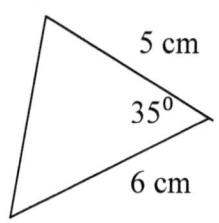

4.

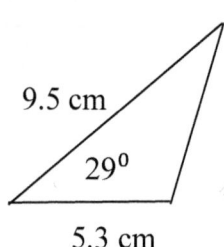

5.

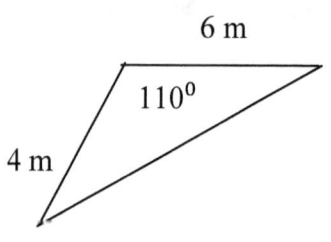

6.

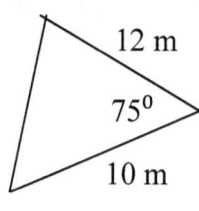

In Exercises 7 – 12, find the length of the side or size of the angle marked x in each of the following triangles, using the given area.

7.

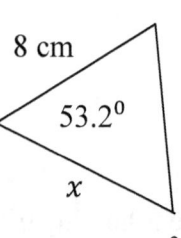

Area = 32 cm^2

8.

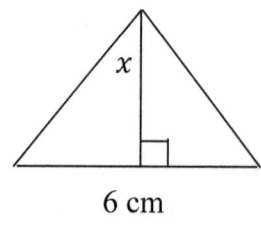

Area = 15 cm^2

9.

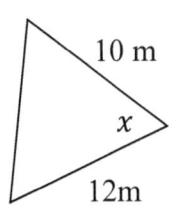

Area = 52 m^2

10.

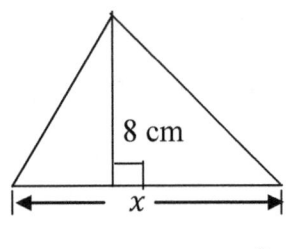

Area = 48 cm^2

11.

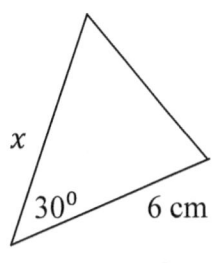

Area = 12 cm^2

12.

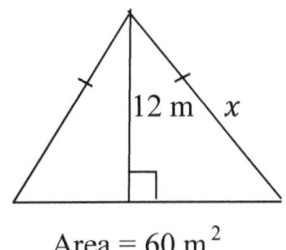

Area = 60 m^2

8.2 Quadrilaterals

Rectangles and Squares

A rectangle is a quadrilateral with four right angles

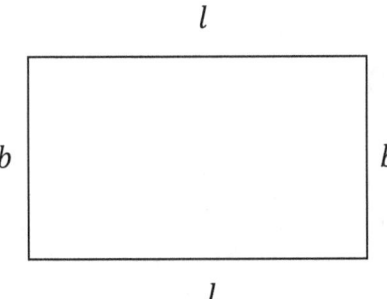

Figure 8.3

Figure 8.3 shows a rectangle of length l and width b. The area, A is given by the formula $A = lb$

The perimeter, P, is $P = 2l + 2b$

$$= 2(l + b).$$

Example

The diagram below is a rectangle of length 5 cm and width 3 cm. Find its area and perimeter.

3 cm

5 cm

$A = lb$

$= 3 \times 5$

$= 15$

The area is 15 cm^{-2}

$Perimeter = 2(l + b)$

$= 2(5 + 3)$

$= 16$

The perimeter is 16 cm

Try this 4

Find the area and perimeter of a rectangle of length 12 cm and width 7 cm.

You may recall that a square has all its sides equal. If the length of a side of a square is l, then the area, A, is $A = l^2$ and the perimeter, P, is $P = 4l$.

Exercise 8.2(a)

1. Find the perimeter and area of the rectangles with the following length and width

(a) 6 cm; 5 cm   (b) 8 m; 7 m

(c) 15 mm; 9 mm   (d) 5.8 m; 4.2 m

2. Find the perimeter and area of the squares with the following sides

(a) 8 cm   (b) 15 cm   (c) 3.5 m   (d) 7.2 cm

3. The area of a rectangle is 20 cm^2. If it is 4 cm wide, find the length of the rectangle.

4. The area of a rectangle is 72 cm^2. If its length is 8 cm, find the width of the rectangle.

5. The length of a rectangle is 5 times its width. If its area is 320 m^2, find the length of the rectangle.

6. A rectangle is 4 cm longer than it is wide. The area of the rectangle is 117 m^2. Find the length and width of the rectangle.

7. A rectangle has width 5 centimetres. If the diagonal is 13 cm long, find its area.

8. A rectangle has length 15 cm. If the perimeter is 54 cm, find its area.

9. Find the perimeter and area of a square that has diagonal 18 cm.

10. If the length of each side of a square is doubled, the area of the square is increased by 675 m². Find the length of the original square.

Parallelograms and Rhombus

A parallelogram is a quadrilateral with both pairs of opposite sides parallel and equal. Any side of a parallelogram can be called a base. For each base there is a corresponding height (or altitude). The height or altitude is the length of a line segment drawn from one base perpendicular to the other base.

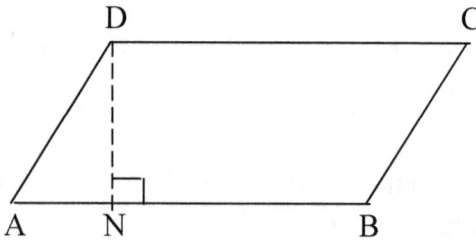

Figure 8.4

Figure 8.4 shows a parallelogram ABCD of height DN. DNA is cut off and fitted at the other end to form the rectangle shown in Figure 8.5

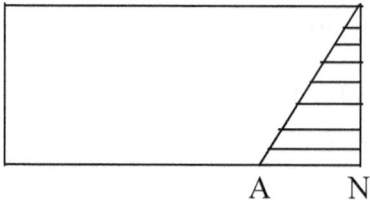

Figure 8.5

The area of the parallelogram is the same as the area of a rectangle that has the same base and heigth. So if a parallelogram has an area, A, a base b and heigth, h then $A = bh$.

A rhombus is a parallelogram that has four sides of equal lenght. To find the area of a rhombus, multiply the lengths of the two diagonals and divide by 2.

Example

Find the area of the parallelogram shown below

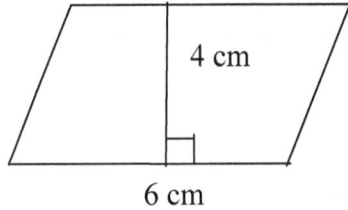

4 cm

6 cm

$A = bh$

$= 6 \times 4$

$= 24$

The area is 24 cm^{-2}

Try this 5

Find the area of the parallelogram shown below

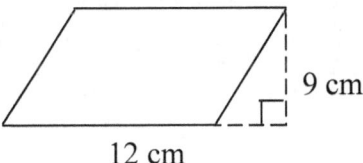

9 cm

12 cm

Example 8.2(b)

In Exercises 1 – 3, calculate the area of each parallelogram

1.   2.   3.

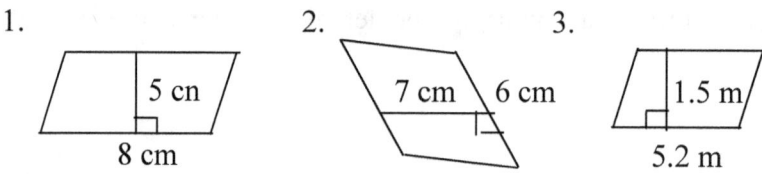

In Exercises 4 – 6, find the lengths of the side marked x in each of the following parallelograms, using the given area

4.   5.   6.

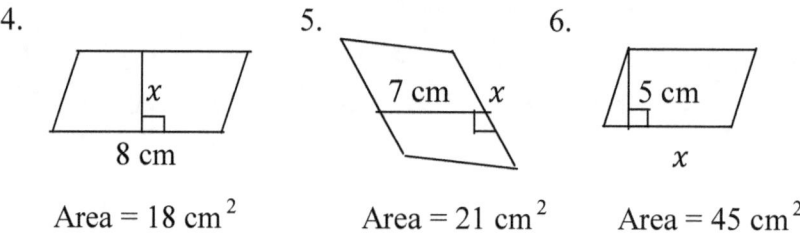

7. The area and length of the base of a parallelogram are 36 m² and 9 m respectively. Find the height of the parallelogram.

8. The areas of a rectangle and a parallelogram are equal. The rectangle has a length of 8 metres and a width of 6 metres. If the parallelogram has a base of 12 metres, find its height.

9. The diagonals of a rhombus are 19 and 12 centimetres long. Find the area of the rhombus.

10. A rhombus has a perimeter of 52 centimetres and a diagonal 24 centimetres long. Find the area of the rhombus.

Area of parallelogram given two sides and included angle

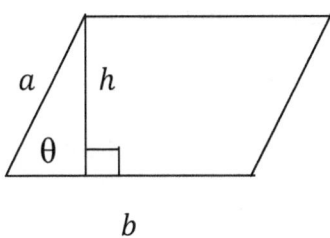

Figure 8.6

Figure 8.6 shows a parallelogram with sides a and b, and an included angle $\theta°$. You can see from the diagram that

$$\frac{h}{a} = \sin \theta°$$

so $h = a \sin \theta°$

Hence, the area, A of the parallelogram is $A = ab \sin \theta°$

Example

Find the area of the parallelogram shown below

$A = ab \sin \theta$

$\quad = 4 \times 5 \times \sin 30°$

$\quad = 10$

The area is 10 cm^2

Try this 6

Find the area of the parallelogram shown below

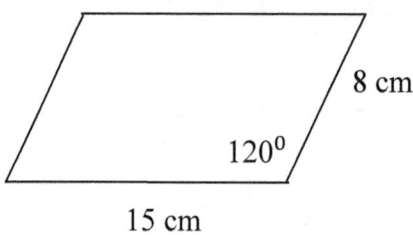

Example 8.2(c)

In Exercises 1 – 3, calculate the area of each parallelogram

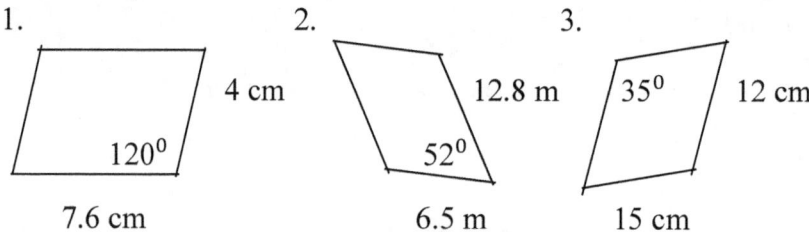

1.   2.   3.

In Exercises 4 – 6, find the length of the side or size of the angle marked x in each of the following parallelograms, using the given area

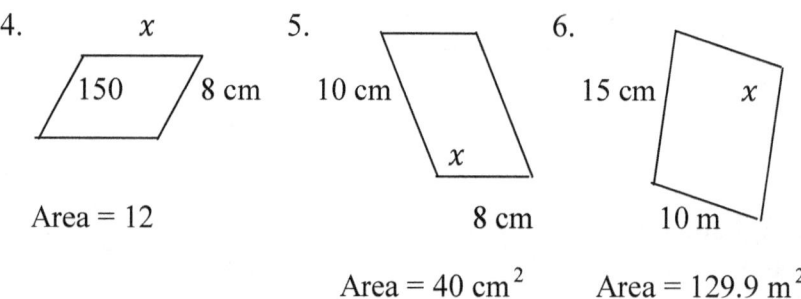

4. x 5.   6.

The Trapezium

A trapezium is a quadrilateral with only one pair of parallel sides. The height of the trapezium is the length of the line segment drawn from one of the parallel sides perpendicular to the other parallel side.

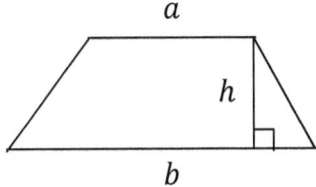

Figure 8.7

The area, A, of the trapezium shown in Figure 8.7 is given by the formula

$$A = \frac{1}{2}(a + b)h$$

where a and b are the length of the two parallel sides and h the heigth.

Example

Find the area of the trapzium shown below

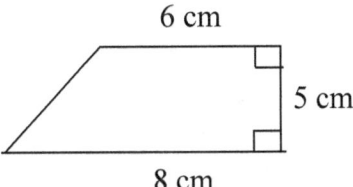

$$A = \frac{1}{2}(a + b)h$$

$$= \frac{1}{2}(6 + 8) \times 5$$

$= 35$

The area is 35 cm $^{-2}$

Try this 7

Find the area of the trapezium shown below

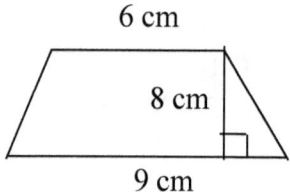

6 cm

8 cm

9 cm

Exercise8.2(d)

In Exercises 1 – 3, calculate the area of each trapezium

1.   2.   3.

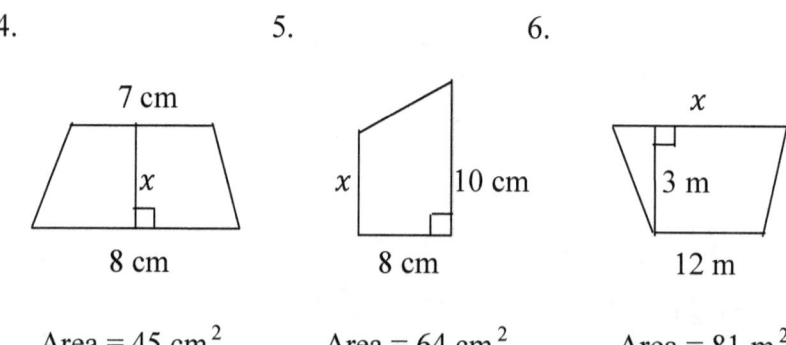

5 cm

4 cm

6 cm

1.2 m   1.5 m

0.9 m

12 cm

7 cm

8 cm

In Exercises 4 – 6, find the length of the side marked x in each case using the given area

4.   5.   6.

7 cm

x

8 cm

x 10 cm

8 cm

x

3 m

12 m

Area = 45 cm ^{2} Area = 64 cm ^{2} Area = 81 m ^{2}

7. A trapezium has an area of 997.5 cm². If the height measures 21 cm and one base measures 40 cm, find the length of the other base.

8. The area of an isosceles trapezium is 36 cm². The perimeter is 28 cm. If the non parallel sides are 5 cm long, find the height of the trapezium.

9.

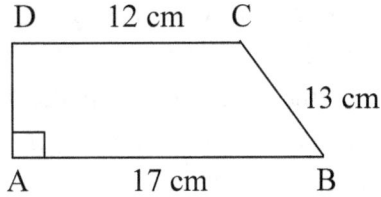

The diagram above shows a trapezium ABCD with AB//DC. AB = 17 cm, BC = 13 cm, DC = 12 cm and $\angle BAD = 90°$. Calculate the area and perimeter of the trapezium.

10.

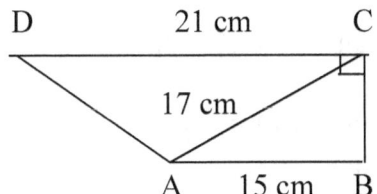

The diagram above shows a trapezium ABCD with AB//DC, AB = 15 cm, AC = 17 cm and DC = 21 cm and $\angle BCD = 90°$. Calculate the area and perimeter of the trapezium.

11.

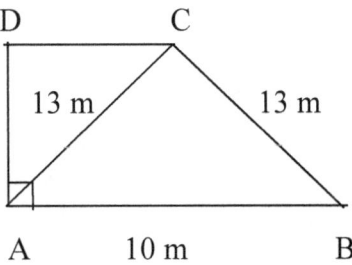

The diagram above shows a trapezium with AB//DC with AB = 10 m, BC = AC = 13 m and $\angle BAD = 90°$. Calculate the area and perimeter of the trapezium.

8.3 Circles

A circle is defined as the set of all points in a plane equidistant from a fixed point called the centre of the circle. A line segment from the centre to any point on the circle is called the radius (plural radii). Any line segment from a point on the circle through the centre to another point on the circle is called a diameter. The length of a diameter is twice the length of the radius of the circle.

The Circumference of a Circle

The perimeter of a circle is called the circumference. The length of the circumference, C, is given by the formula

$C = \pi d$, where d is the diameter or $C = 2\pi r$, where r is the radius.

The number π is $\frac{22}{7}$ and can be approximated by 3.142. If the radius or diameter is exactly divisible by 7, it is often more convenient to use $\pi = \frac{22}{7}$.

Examples

Calculate the circumference of a circle when the diameter is 21 cm.

$C = \pi d,.$

$= \frac{22}{7} \times 21$

$= 66$

The circumference is 66 cm

Try this 8

Calculate the circumference of a circle when (a) the diameter is 63 cm and (b) the radius is 3.5 m

Calculate the radius of a circle when the circumferenc is 20 cm.

If the radius is r cm, then the circumference is $2\pi r$ cm. Hence

$$2\pi r = 20$$

$$r = \frac{20}{2\pi}$$

$$= 3.2$$

The radius is 3.2 cm

Try this 9

Calculate the radius of a circle when the circumference is 35.2 m

Exercise 8.3(a)

1. Calculate the circumference of a circle when the diameter is:

(a) 28 cm   (b) 6.3 m   (c) 15 m   (d) 8.2 km

2. Calculate the circumference of a circle when the radius is:

(a) 2.1 m   (b) 14.7 mm   (c) 6 cm   (d) 45 km

3. Calculate diameter of a circle when the circumference is:

(a) 22 cm   (b) 7.7 km   (c) 14.3 m   (d) 35 cm

4. Calculate the radius of a circle when the circumference is:

(a) 6.4 km   (b) 8 m   (c) 10 cm   (d) 31.4 m

5. The wheel on a boy's bicycle has a radius of 28 cm. How many revolution the wheel make while he travels 352 m.

6. A bicycle wheel diameter 6.3 cm makes a complete revolution in 5 seconds. What distance does it move in 30 minutes.

7.

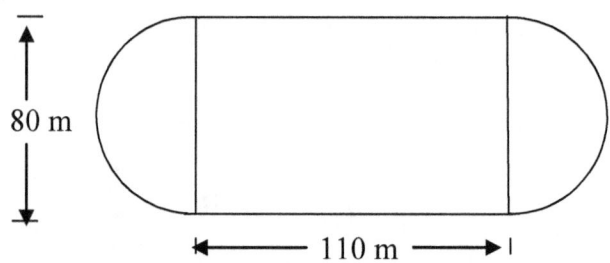

The diagram of a track and a football field is showm above. Find, correct to the nearest unit, the distance round the track.

8. A racing track consists of two straight sides, each 142 m long, connected by two semi -circles, each of diameter 42 m. How many laps do an althlete do in a 5-kilometer race?

9. A piece of wire 3.6 m long is bent into the form of a semicircular arc and its diameter. Find the radius.

10. A bucket is raised from the botton of a well 26.4 m deep by a rope wound on an axle of radius 3 cm. How many turns of the axle are required to bring the bucket up to the top

[Neglect the thickness of the rope]

11. A track is pulled up a slope by a rope wound on a drum which diameter is 21 cm. If the drum was turned 25 times to pull the truck up the slope, how long is the slope?

The Length of an Arc

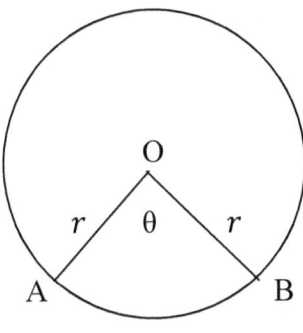

Figure 8.8

Figure 8.8 shows a circle radius r, and centre O. The minor arc AB subtends an angle $\theta°$ at the centre of the circle. The length of the arc is a fraction of the circumference of the circle. The fraction of the circumference that the angle θ represnts is $\frac{\theta}{360}$. Hence, the length of the arc AB is

$$length\ of\ arc = \frac{\theta}{360} \times 2\pi r$$

Example

Find the length of the arc AB in the cirle shown below.

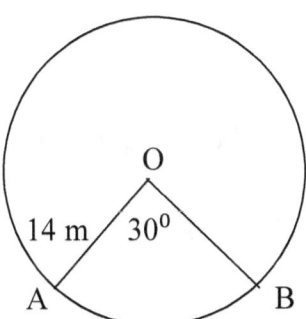

Length of arc $= \dfrac{\theta}{360} \times 2\pi r$

$$= \dfrac{30}{360} \times 2 \times \dfrac{22}{7} \times 14$$

$$= 7\dfrac{1}{3}$$

The length of arc AB is $7\dfrac{1}{3}$ cm

Try this 10

Find the length of an arc of circle radius 28 cm that subtend angle 120^0 at the centre.

Example

An arc of length 4.4 cm is taken from a circle of radius 3 cm. What angle does the arc subtend?

Lenngth of arc $= \dfrac{\theta}{360} \times 2\pi r$

So, $\dfrac{\theta}{360} \times 2 \times \dfrac{22}{7} \times 3 = 4.4$

$$\theta = \dfrac{4.4 \times 360 \times 7}{2 \times 22 \times 3}$$

$$= 84°$$

Try this 11

An arc of length 11 cm is taken from a circle of radius 14 cm. What angle does the arc subtend?

Exercise 8.3(b)

1. Find lengths of the arcs taken from circles with the following radii and sector angles

(a) 3 m; 90⁰   (b) 120 mm; 45⁰

(c) 2.5 m; 180⁰   (d) 12.5 cm; 144⁰

2. Find the sectors angles subtended by arcs of circles with the following arc lengths and radii.

(a) 13 cm; 10 cm   (b) 28.3 cm; 15 cm   (c) 2.5 m; 2.4 m

3. An arc of angle 45⁰ has length 25 m. What is the radius of the circle from which it is taken?

4. An arc of angle 76⁰ has length 15.9 cm. What is the radius of the circle from which it is taken?

5. Find the perimeter of the sectors of the circles with the following radii and sector angles

(a) 7 cm; 120⁰   (b) 10 cm; 45⁰   (c) 21 m; 60⁰

6. A piece of wire is bent into the arc of a circle with radius 25 cm. If the arc subtends an angle of 13⁰ at the centre, what is the length of wire?

7. How far will the tip of the long hand of a clock, of radius 11 cm, travel in 45 minutes?

The Area of a Circle

The area, A, of a circle is given by the formula

$$A = \pi r^2$$

where r is the radius.

If the diameter, d, is given instead of the radius, we have

$$A = \pi \left(\frac{d}{2}\right)^2 = \frac{1}{4}\pi d^2$$

Examples

Calculate the area of a circle when the diameter is 28 cm.

The diameter is twice the radius of the circle, so $r = \frac{28}{2} = 14$ cm.

$$A = \pi r^2$$

$$= \frac{22}{7} \times 14^2$$

$$= 616$$

The area is 616 cm ^{2}

Try this 12

Calculate the area of a circle when (a) the radius is 5 cm and (b) the diameter is16 m

The area of a circle is 154 cm^2, calculate its radius.

If the radius of the circle is r, then

$$A = \pi r^2.$$

So, $\frac{22}{7} \times r^2 = 154$

$$r^2 = \frac{154 \times 7}{22}$$

$$= 49$$

$$r = 7$$

The radius is 7 cm

Try this 13

The area of a circle is 24.64 cm², calculate its radius.

Exercise 8.3(c)

In each of the following exercises, round answers to the nearest tenth when necessary.

1. Calculate the area of a circle when the radius is:

(a) 2.5 m   (b) 5.6 cm

(c) 49 mm   (d) 1.6 km

2. Calculate the radius of a circle when the area is:

(a) 198 cm²   (b) 24.2 m²

(c) 500 mm²   (d) 80 cm²

3. A running truck consists of two parallel straight sides 20 m apart joined by two semi- circular areas. If the straight sides are each 50 m long, find the length of the track and the area it encloses.

4.

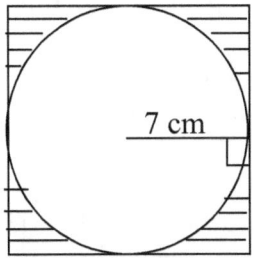

7 cm

The diagram shows a circle of radius 7 cm inscribed in a square. Find the area of the shaded region.

5.

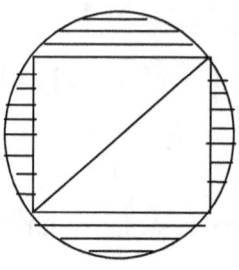

The diagram shows a square of side 6 cm inscribed in a circle. Find the area of the shaded region.

6.

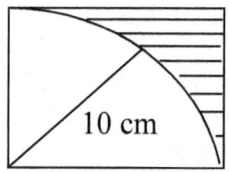

10 cm

The diagram shows a quarter-circle inscribed in a square. Find the area of the shaded region.

7. What is the area of grass a goat can eat if it is fastened to a peg in the ground by a rope 2.1 m long?

8. A certain tin of paint is sufficient to paint 75 m². What is the radius of the largest circle that can be painted with this tin of paint?

9. Two circular lead discs of radii 3 cm and 4 cm respectively are melted and cast into another circular disc of the same thickness. Find the radius of this disc.

10. A running truck consists of two parallel straight sides 80 m apart joined by two semicircular areas. The straight sides are each 110 m long. If it cost 25 Gp to grass a square metre, how much would it cost to grass the area enclosed by the track.

11. Find the area of each shaded region. O is the centre of the circle.

(a)   (b)

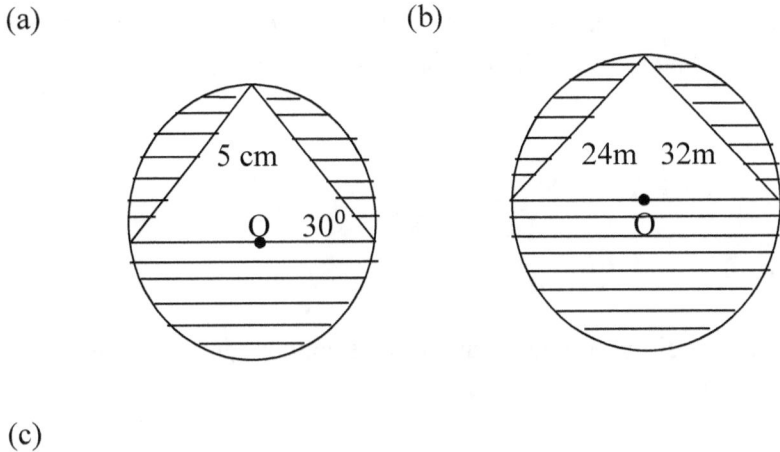

(c)

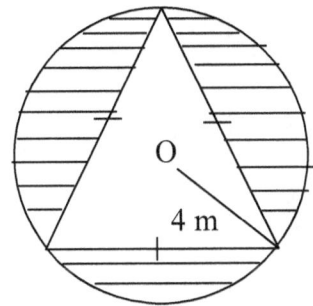

The Area between Two Circles

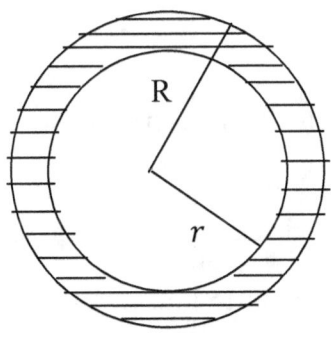

Figure 8.9

Figure 8.9 shows two concentric circles with radii of R and r. The area of the shaded part of the circle is called annulus or ring. You can find the area of the ring by subtracting the area of the inner circle from that of the outer circle. Hence, the area A of the ring is

$$A = \pi R^2 - \pi r^2$$

$$= \pi(R^2 - r^2)$$

Example

Find the area of a circular ring formed by two concentric circles with radii of 6 cm and 8 cm respectively

$$A = \pi(R^2 - r^2)$$

$$= \pi(8^2 - 6^2)$$

$$= \frac{22}{7}(64 - 36)$$

$$= 88$$

The area of ring is 88 cm^2

Try this 14

Find the area of a circular ring formed by two concentric circles with radius 7 cm and 12 cm respectively.

Exercise 8.3(d)

In each of the following exercises, round answers to the nearest tenth when necessary.

1. Find the area of two concentric circles with the following radii

(a) 4 cm; 3 cm

(b) 22 mm; 15 mm

(c) 12 m; 8 m

(d) 2.5 m; 2.4 m

2. A circular sheet of radius 9 cm has a circlar hole cut in it of radius 5 cm. Find the area of metal remaining.

3. Find the area of a path 1.1 m wide surrounding a circular pond of diameter 3.8 m.

4. A round about on a main road is of diameter 140 m and is surrounded by a road 20 m wide all the way round. Find the area of the road surrounding the round about. [Take $\pi = 3.142$]

5. A metal washer has an outside radius 6.8 mm and inside radius 3.3 mm. Find the area of the metal. [Take $\pi = \frac{22}{7}$]

6. A circular mirror is suurounded by a wooden frame 1 cm wide. If the radius of the mirror is 7 cm find the area of the frame.

7. A circlar pond of diameter 28 cm is surrounded by a path 7 cm wide. Find the area of the path.

The Area of a Sector

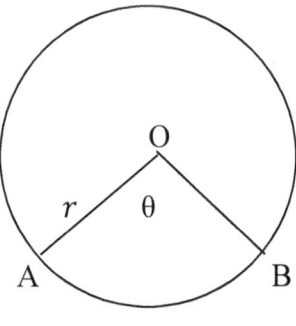

Figure 8.10

Figure 8.10 shows a circle centre O, with radius r. The minor arc AB subtends an angle θ at the centre. The area of the sector OAB is a certain fraction of the area of the circle. You may recall that the fraction of the sector represented by angle θ is $\frac{\theta}{360}$. Hence the area, A, of the sector is

$$A = \frac{\theta}{360} \times \pi r^2$$

Examples

A sector of angle 72^0 is taken from a circle of radius 14 cm. Find the area of the sector.

$$A = \frac{\theta}{360} \pi r^2$$

$$= \frac{72}{360} \times \frac{22}{7} \times 14^2$$

$$= 123.2$$

The area of the sector is 123.2 cm ^{2}

Try this 15

Find the area of a sector of a circle of radius 8 cm which subtend an angle of 45^0 at the centre.

A sector of area 154 cm^2 is taken from a circle of radius 21 cm. What is the angle of the sector?

$$A = \frac{\theta}{360} \pi r^2.$$

So, $\frac{\theta}{360} \times \frac{22}{7} \times 21^2 = 154$

$$\theta = \frac{154 \times 360 \times 7}{22 \times 21^2}$$

$$= 40$$

The angle of the sector is 40^0

Try this 16

A sector of area 20.8 cm^2 is taken from a circle of radius 6.3 cm. Find the angle of the sector.

Exercise 8.3(e)

1. Find the areas of the sectors of the circles with the following radii and sector angles

(a) 70 mm; 36^0 (b) 21 cm; 120^0

(c) 2 m; 135^0 (d) 6 cm; 108^0

2. The area of a sector of angle 30^0 is 25 cm^2. Calculate the radius of the circle?

3. A sector of angle 140^0 has area 792 cm^2. Calculate the radius of the circle?

4. A sector of angle 10^0 has area 20 cm^2. Calculate the radius of the circle from which it is taken?

5. A sector of circle radius 6 m has an area 19.8 m^2. Find the sector angle.

6. Find the area of the shaded segment shown in each diagram below

(a)   (b)

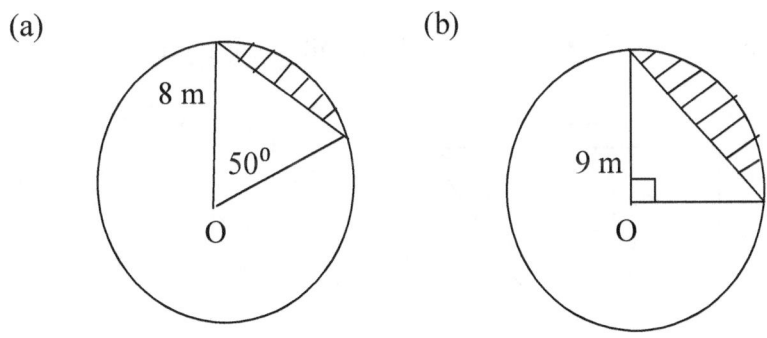

7. A chord subtends 70^0 at the centre of a circle of radius 14 cm. Find correct to one decimal place, the area of the segment cut off by the chord.

8. A chord subtends 60^0 at the centre of a circle and the area of the segment cut off is 12 cm². Find the radius of the circle.

Review exercise 8

1. Find the area of each of the following triangles:

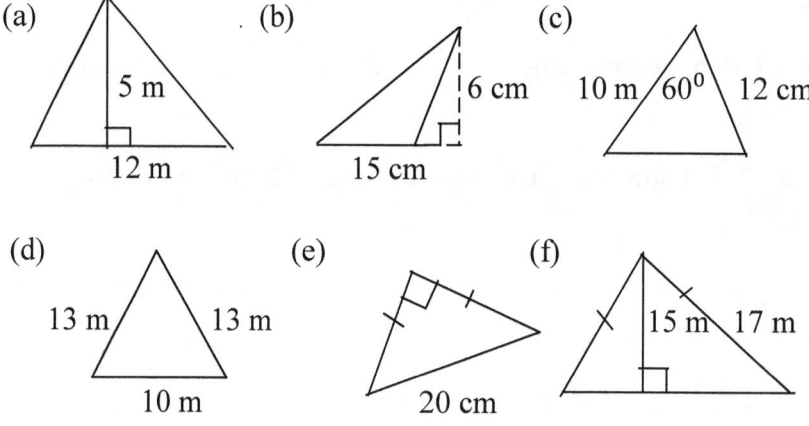

(a)   5 m   12 m

(b)   6 cm   15 cm

(c)   10 m 60^0 12 cm

(d)   13 m   13 m   10 m

(e)   20 cm

(f)   15 m 17 m

2. In triangle ABC, AB = 6 cm, BC = 8 cm and AC = 10 cm and $\angle ABC = 90°$. Find the length of the altitude from the vertex B to $\overline{AC}$

3. Find the area and perimeter of each of the following figures:

(a)

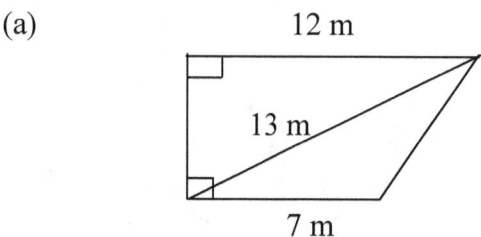

12 m

13 m

7 m

(b)

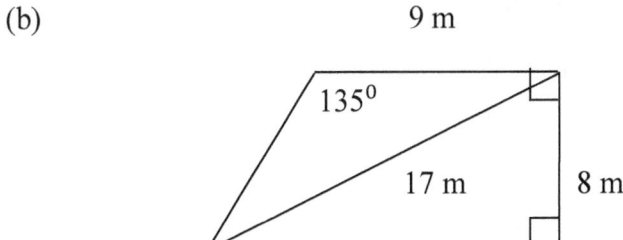

4. The area of a triangle is 88 cm ^{2}. If the height is 16 cm, what is the length of the base?

5. If the area of a rectangle is 243 m ^{2} and the length is 27 m, what is the width of the rectangle?

6. If the width of a rectangle is 29 m and the length is 37 m, what is the area of the rectangle?

7. What is the length of a rectangle with perimeter 110 cm and width 26 cm?

8. What is the area of a square with perimeter 56 cm?

9. If the width of a rectangle is 2.4 m and the area is 8.4 square metre, find the perimeter of the rectangle?

10. A rectangle is 3 cm longer than it is wide. If it has a perimeter of 24 cm, find the area of the rectangle.

11. Calculate the area of a square when the perimeter is 28 cm.

12. Find the area of each of the following parallelograms.

(a)   (b)   (c)

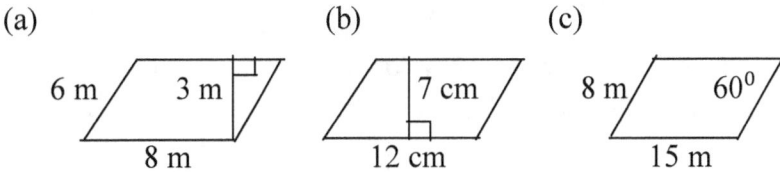

13. Find the area of each trapezium

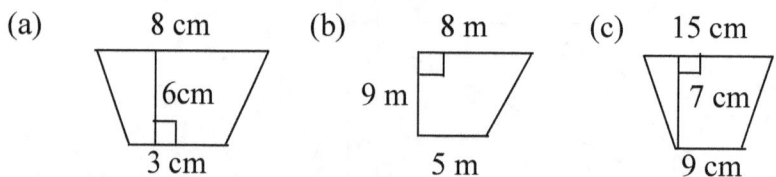

(a)   8 cm   (b)   8 m   (c)   15 cm
      6cm   9 m   7 cm
      3 cm   5 m   9 cm

14. Calculate the circumference of a circle when the diameter is:

(a) 2.8 m   (b) 6.4 cm   (c) 12 mm

15. Calculate the circumference of a circle when the radius is:

(a) 21 cm   (b) 4.5 m   (c) 7.2 cm

16. An arc of circle radius 5 cm subtend an angle 50^0 at the centre. Find the length of the arc.

17. An arc of length 30 cm is taken from a circle of radius 20 cm. Find the angle subtended by the arc.

18. A sector of angle 60^0 is taken from a circle of radius 21 cm. Find the perimeter of the sector. [Take $\pi = \frac{22}{7}$]

19. Calculate the area of a circle when the radius is:

(a) 4.9 m   (b) 5.6 cm   (c) 7.2 cm

20. Find the area of a circular ring formed by two concentric circles with radii 8 cm and 13 cm respectively.

21. Find the circumference and area of a circle inscribed in a square which sides are 6 metres long

22. The diagonal of a square is 8 cm long. Find the circumference and the area of a circle inscribed in the square.

23.

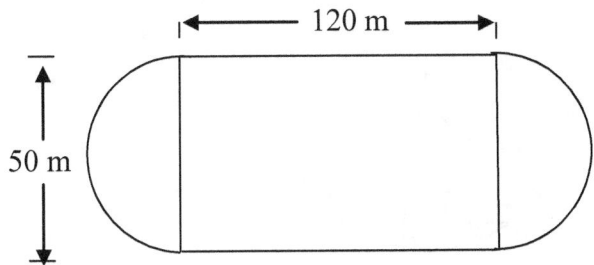

The diagram shows a running track consisting of two parallel straight lines of length 120 m, separated by semi- circular ends of diameter 50 m. Find the length of the track and the area it encloses.

24

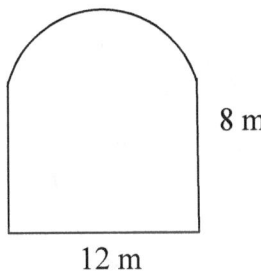

8 m

12 m

The figure above is formed from a semi - circle and a rectangle. Find the area and perimeter of the figure.

25. A sector of angle 75^0 is taken from a circle of radius 6 cm. Find the area of the sector.

26. A sector of area 231 cm^2 is taken from a circle of radius 14 cm. Find the angle of the sector.

27. The area of a sector of circle radius 5.4 cm is 21.6 cm^2. Find the length of the arc.

28. A chord subtends 86^0 at the centre of a circle and the area of the segment cut off is 27 cm^2. Find the radius of the circle.

29.

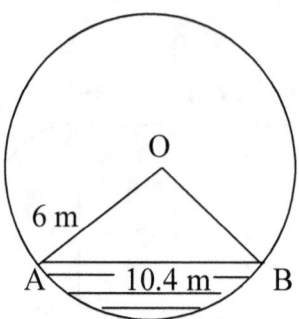

The diagram above shows a circle, centre O, radius 6 m and AB = 10.4 m. Calculate the area of the shaded segment.

30.

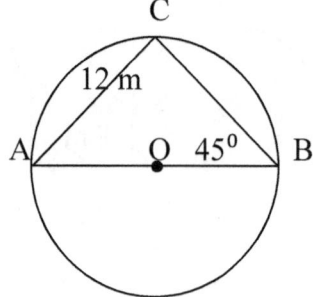

In the diagram, triangle ABC is cut off from the circle, centre O. If AC = 12 m and ∠ABC = 45°, find the area of the remaining part of the circle.

Chapter Test 8

Take this test as you would take a test in a class. After you are done, check your work against the answers in the back of the book.

1. Calculate the area and perimeter of the following shapes. O is the centre of the semicircle.

(a)

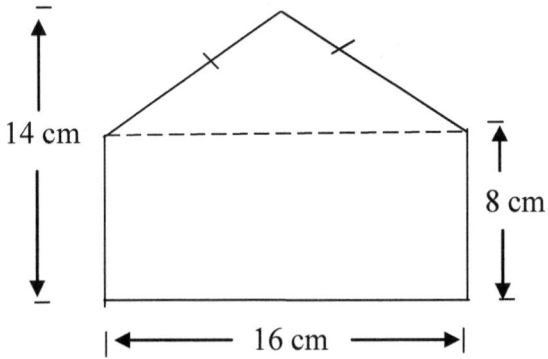

(b)

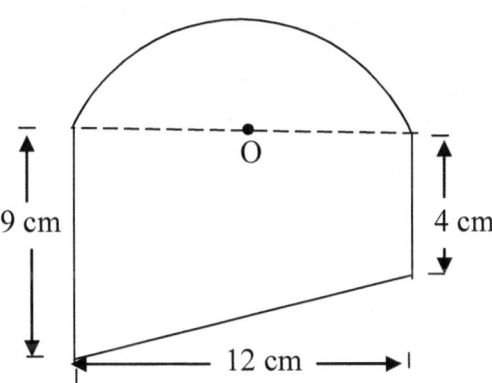

2.

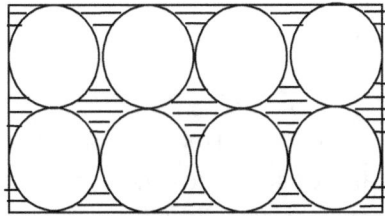

Eight circles are arranged inside a rectangle as shown. If the radius of each circle is 4 centimetres calculate:

(a) the perimeter of the rectangle

(b) the shaded area within the rectangle

3.

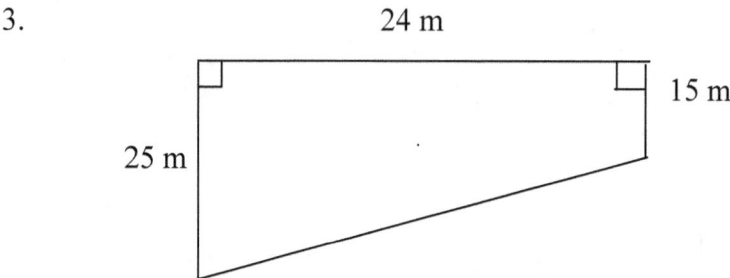

The diagram represents a farmer's field. The farmer decides to put up a fence around the field. If it cost GH¢ 1.50 to put one metre of fence, how much would it cost to fence the whole field?

4. A rectangular garden is 24 m by 32 m. A side walk is built along the inside edges of all four sides. If the garden now has an area of 468 m², how wide is the side walk?

5.

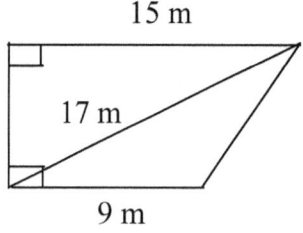

The diagram represents a school farm. The school wants to plant seed in the farm. If one kilogram of seeds covers 12 m², how many kilogram of seed does it take to cover the whole farm

6. Two wire hoop of diameters 3 cm and 4 cm, are broken and the wire used to make another larger circular hoop'. What is the diameter of this hoop?

7.

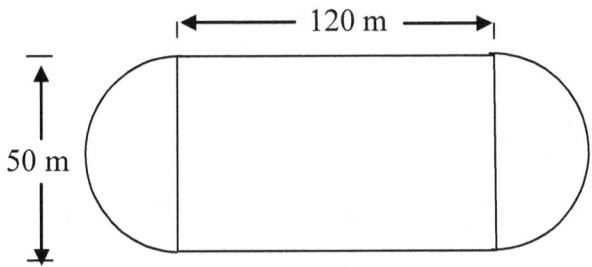

A diagram of a racing track at a school is shown above.

(a) Calculate the total area of the track

(b) How many laps do an aithlete do in a 5- kilometer race?

8. A circular sheet of radius 2.5 cm has a circular hole cut in it of radius 2 cm. Find the area of the remaining metal.

9. O is the centre of the circle which has a radius of 5.4 cm.

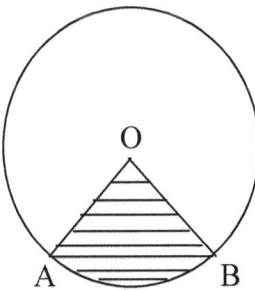

The area of the shaded sector OAB is 21.6 cm².

(a) Find the length of the minor arc AB

(b) Find the perimeter of the sector OAB.

[Take $\pi = \frac{22}{7}$]

10

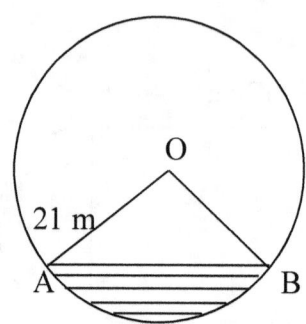

The diagram shows a circle of centre O, and radius 21 m. If the perimeter of the minor sector OAB is 64 cm, calculate correct to one decimal place, the

(a) length of the chord AB

(b) area of the shaded segment

9

Surface Areas and Volumes of Solids

9.1 Prism

A prism is a solid that has two faces that are parallel and congruent, called the bases. The other faces that are not bases are called lateral faces. A segment that is the intersection of two faces is called an edge. A point that is the intersection of three or more edges is called a vertex (plural vertices).

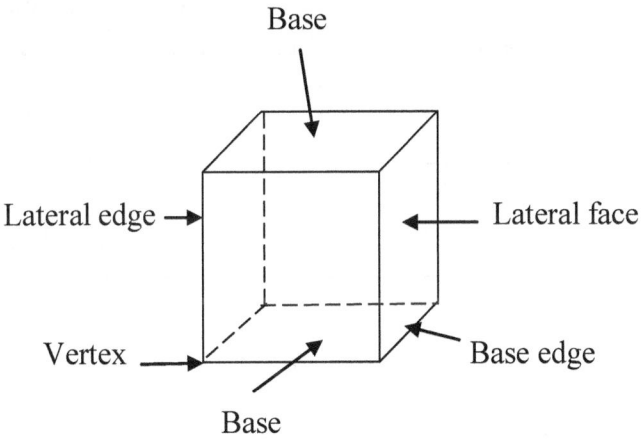

Figure 9.1

The height h of a prism is the length of any perpendicular segment drawn from a point on one base to the other base. If the lateral edges of a prism are perpendicular to the bases, then the prism is called a right prism. A prism is named by the shape of its bases. For example, a prism that has a triangle as its base is called a triangular prism.

The Surface Area of a Prism

The surface area of a prism is the sum of the areas of its faces.

Area of Lateral Faces

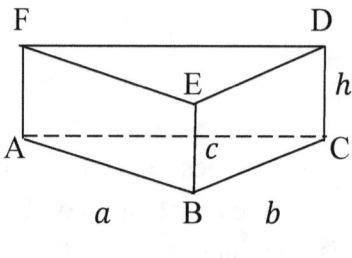

Figure 9.2

Figure 9.2 shows a prism with triangular base with sides a, b and c. The height of the prism is h. Each lateral face is a rectangle. The prism can be spread out into one continuous rectangle as shown in Figure 9.3

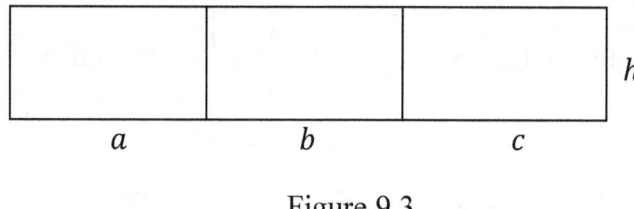

Figure 9.3

The lateral area, A, of the prism is the sum of the areas of its lateral faces. Hence

$$A = ah + bh + ch$$

$$= (a + b + c)h$$

Recall that the perimeter, P, of the triangle is $a + b + c$. Generally, you can calculate the lateral area of a prism by multiplying the height, h, of the prism by the perimeter, P, of its base i.e. $A = Ph$.

Surface Area

The surface area of a prism is the sum of the area of each lateral face and the area of each base. Since the bases are congruent, they have the same area.

Examples

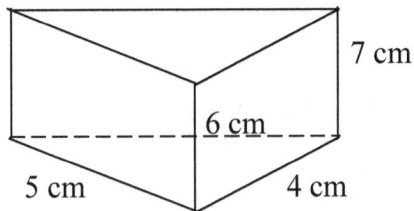

The diagram shows a prism with a triangular base with sides 5 cm, 4 cm and 6 cm. The height of the prism is 7 cm. Find the area of the lateral faces

$$Area\ of\ lateral\ face = (a + b + c)h$$

$$= (5 + 4 + 6) \times 7$$

$$= 15 \times 7$$

$$= 105\ cm^2$$

Try this 1

A right prism stands on a triangular base with sides 7 cm, 8 cm and 10 cm. If the height of the prism is 12 cm, find the area of the lateral faces.

A right prism stands on a triangular base with sides 3 cm, 4 cm and 5 cm. If the height of the prism is 6 cm, calculate the surface area.

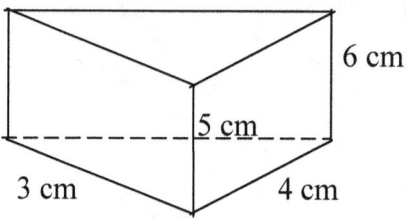

$Area\ of\ lateral\ face = (3 + 4 + 5) \times 6$

$$= 12 \times 6$$

$$= 72\ cm^2$$

$Area\ of\ a\ base = \frac{1}{2} \times 3 \times 4$

$$= 6\ cm^2$$

$Surface\ area = 72 + 2(6)$

$$= 84\ cm^2$$

Try this 2

A right prism stands on a rectangular base with sides 5 cm and 6 cm. The height is 8 cm. Calculate the surface area.

The Volume of a Prism

Recall that a prism has two congruent and parallel faces, called the bases of the prism. The volume of any prism can be found by multiplying the area of one of the bases by its height. If a right prism has a volume, V, a base area, A, and height, h then

$$V = Ah$$

Example

The base of a right prism is a square with side 5 cm. If the height of the prism is 6 cm, find the volume of the prism.

$Area\ of\ base = 5^2$

$$= 25\ cm^2$$

$Volume\ of\ prism = Ah$

$$= 25 \times 6$$

$$= 150\ cm^3$$

Try this 3

Find the volume of a right prism with height 15 cm, and a right triangular base with sides 5 cm, 12 cm and 13 cm.

Exercise 9.1

1. A right prism stands on a triangular base that has sides 15 cm, 12 cm and 9 cm. If the height of the prism is 10 cm calculate the area of the lateral faces.

2. A right prism stands on a triangular base. The triangular bases are equilateral with sides 5 cm. If the height of the prism is 12 cm calculate the area of the lateral faces.

3. Calculate the surface area of each of the following prisms. Round off your answer to the nearest tenth if necessary.

(a)   (b)

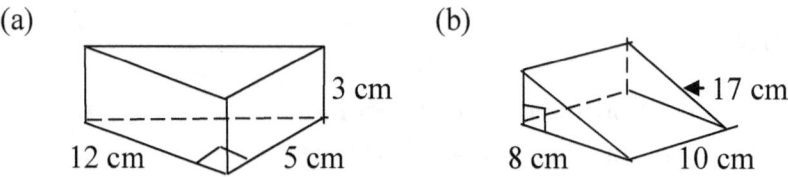

(c)

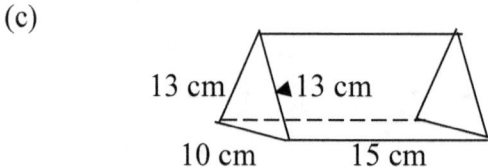

4. The base of a right prism is a rectangle that has length 20 cm and width 5 cm. If the height of the prism is 3 cm, calculate the surface area of the prism

5. The base of a prism is a square of side 10 cm. The prism is 6 cm long. Calculate the total surface area of the prism.

6. The base of a right prism is a rhombus side 5 cm. If the height of the prism is 8 cm, calculate the surface area of the prism.

7. The surface area of a cube is 864 cm^2. Calculate the length of the lateral edge of the cube.

8. The surface area of a rectangular prism is 256 cm^2. If the sides of the base are 7 cm and 9 cm, calculate the height of the prism

9. Calculate the volume of each of the following prisms. Round off your answer to the nearest tenth if necessary.

(a)   (b)   10 cm

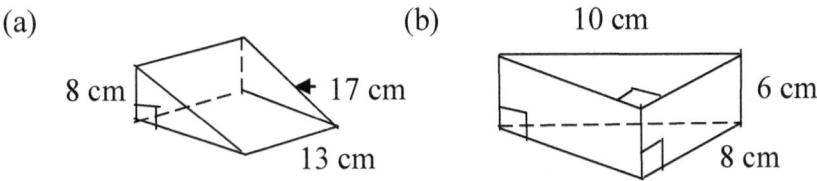

(c)

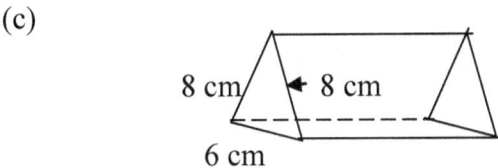

10. The volume of a right rectangular prism is 1153 cm^3, and the area of each base is 64 cm^2. Calculate the length of the lateral edge of the prism.

11. A right prism stands on a triangular base that has two sides AB = AC = 5 cm. If the height and surface area are 8 cm and 144 cm^2 respectively calculate the volume of the prism.

12. A right prism stands on a triangular base that has sides 5 cm, 4 cm and 3 cm. If the height of the prism is 21 cm, calculate the volume of the prism.

13. The base of a right prism is a trapezium which parallel sides are 12 cm and 13 cm. The distance between the parallel sides is 8 cm. Calculate the volume of the prism when the height is 10 cm.

14. The cross-section of a container is a square on a side of 8 cm. If the depth of water after 12 seconds is 3 cm, calculate the rate of flow of water in cm^3 s^{-1}.

15. The length of the base of a right rectangular prism is three times its width. The height is twice its width. If the area of the lateral face is 144 square metres, calculate its volume.

16. Calculate the surface area and volume of each of the following prism. Round off your answer to the nearest tenth if possible.

(a)

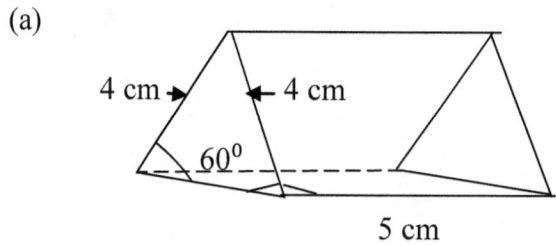

(b)

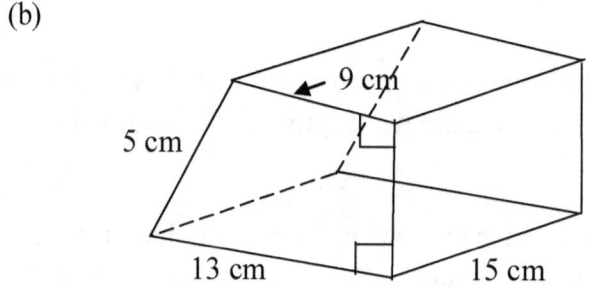

(c)

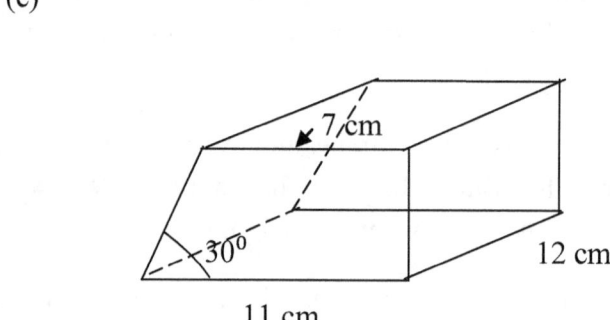

9.2 Cuboids and Cubes

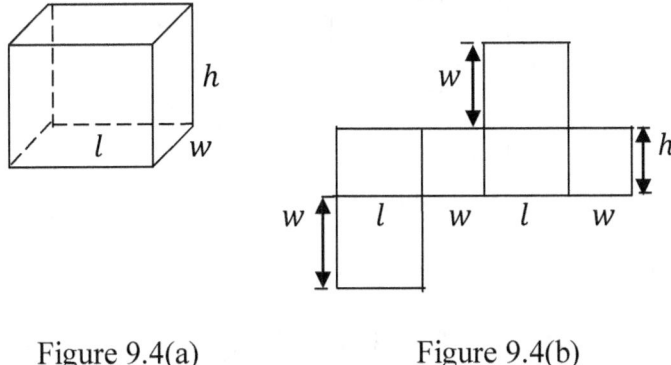

Figure 9.4(a)   Figure 9.4(b)

The right prism shown in Figure 9.4(a) stands on a rectangular base with length l and width w. The height of the prism is h. This prism is called a cuboid. An example of a cuboid is a box. The net of a rectangular cuboid is shown in Figure 9.4(b)

The surface area is made up of three pairs of opposite and equal rectangular faces. The areas of the lateral faces and the two bases are $(2l + 2w)h$ and $2lw$ respectively. Hence, the total surface area A of the cuboid is

$$A = 2lh + 2wh + 2lw$$

A cube is a special case of a cuboid. All the edges of a cube are equal. That is $l = w = h$. The surface area, A of the cube is

$$A = 6l^2$$

Example

A solid box has dimensions 2 cm, 5 cm and 7 cm. Find its surface area

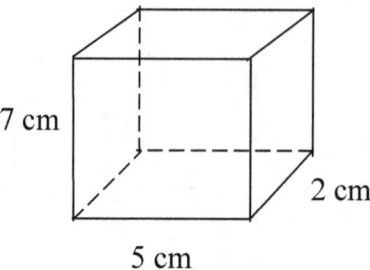

Surface area $= 2lh + 2wh + 2lw$

$$= 2(5)(7) + 2(2)(7) + 2(2)(5)$$

$$= 70 + 28 + 20$$

$$= 118 \ cm^2$$

Try this 4

A solid box has dimensions 5 cm, 6 cm and 9 cm. Find its surface area

The Volume of a Cuboid and a Cube

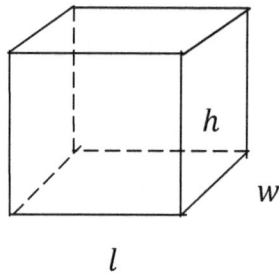

Figure 9.5

Figure 9.5 shows a cuboid length, l, width w and height h. Recall that the volume of a prism is Ah, where A is the area of the base and h, the height of the prism. Hence, the volume, V, of the cuboid is

$$V = lwh$$

If each edge of a cube is l unit, then the volume, V, is

$$V = l^3$$

Example

A solid box has dimensions 3 cm, 7 cm and 8 cm. Find its volume.

$Volume\ of\ box = 3 \times 7 \times 8 = 168\ cm^3$

Try this 5

A solid box has length 5 cm, width 3 cm and height 7 cm. Calculate the volume of the solid.

Exercise 9.2

1. A solid box has dimensions 3 cm, 6 cm and 8 cm. What is its surface area?

2. A rectangular tin can has dimensions 6 cm, 4 cm and 9 cm. Find its volume.

3. Calculate the surface area of a cube if the lateral edge is 8 cm long.

4. Calculate the volume of a cube when the edge is 9 cm

5. A rectangular carton has a base measuring 12 cm by 15 cm. How high must it be to hold 1.8 litres of milk?

6. A container 20 cm by 40 cm and 4 cm deep is filled to the brim with a liquid. If the liquid weighs approximately 0.25 grams per cubic centimetre, what is the approximate weight of liquid in the container?

7. A rectangular water tank 12 cm wide, 15 cm long and 8 cm high leaks. If it loses 24 cm^3 of water every 5 minute when will the tank be empty?

8. What are the dimensions of a solid cube that has volume numerically equal to one- third of its surface area?

9. The surface and base areas of a rectangular block are 214 cm^2 and 42 cm^2 respectively. If the area of one vertical face is 35 cm^2, calculate the length of the edges.

10. The diagonal of a rectangular solid is 17 cm, and the surface area is 552 cm^2. Find the sum of the three dimensions.

11. Find the volume of a cube for which a diagonal of one of its faces measures 12 centimetres.

12. Find the cost of painting the four inner sides and bottom of a tank which measures internally 2.5 m long, 1.2 m wide and 1.5 m deep, at GH¢2.50 per square metre.

13. A rectangular swimming pool has a volume of 16,320 cubic metres, a depth of 8 metres and a length of 85 metres. What is the width of the swimming pool?

14. A rectangular tank holds 4 litres of water and is 10 centimetres deep and 25 centimetres long. What is the width of the tank?

15. A rectangular tank is 50 centimetres long and can hold water up to 16 centimetres deep. The tank was filled to the depth of 6 centimetres in 5 seconds at a rate of 1.5 litres per second. How much water had to be added to fill the tank?

9.3 Cylinders

A cylinder is a right prism that has two circular bases connected by a curved surface.

The Curve Surface Area

Figure 9.6

The rectangular sheet shown in Figure 9.6 can be bent round to form the curved surface area of a cylinder as shown in Figure 9.7.

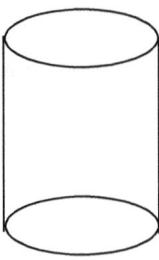

Figure 9.7

You may have notice that the width of the rectangular sheet is the height of the cylinder and the length is equal to the circumference of its base. The curved surface area is the same as the area of the rectangular sheet. If the height of the cylinder is h unit, and the radius of the base is r unit, then the curved surface area, A of the cylinder is

$$A = 2\pi rh$$

The Surface Area

The diagram shows the net of a closed cylinder

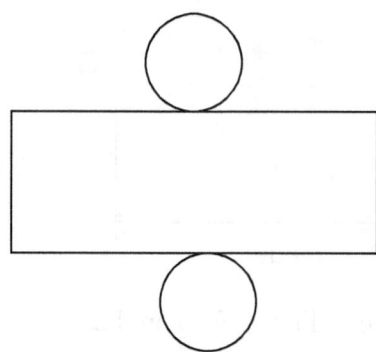

The surface area of the cylinder is the sum of the area of the rectangle and the area of each of the two circles.

Therefore the surface area, A, is

$A = 2\pi rh + 2\pi r^2$

This can be factorised to give

$A = 2\pi r(h + r)$

When the cylinder is closed at one end and opens at the other, the area, A, of the surface is $A = 2\pi rh + \pi r^2$.

Examples

Find the curved surface area of a cylinder radius 3 cm and height 14 cm.

$Curved\ surface\ area = 2\pi rh$

$$= 2 \times \frac{22}{7} \times 3 \times 14$$

$$= 264\ cm^2$$

Try this 6

Calculate the curved surface area of a cylinder radius 5 cm and height 6.3 cm

Calculate the total surface area of a cylinder radius 2.1 cm and height 10 cm.

$$surface\ area = 2\pi rh + 2\pi r^2$$

$$= 2\pi r(h + r)$$

$$= 2 \times \tfrac{22}{7} \times 2.1(10 + 2.1)$$

$$= 159.7\ cm^2$$

Try this 7

Calculate the total surface area of a cylinder radius 6.3 cm and height 15 cm.

The Volume of a Cylinder

Recall that a cylinder is a prism with a circular base so its volume is the product of the area of the base and the height. The area of the base of the cylinder with radius r is πr^2. If the height of the cylinder is h, then the volume, V, of the cylinder is

$$V = \pi r^2 h$$

Example

Find the volume of the cylinder with radius 2 cm and height 7 cm.

$$Volume = \frac{22}{7} \times 2^2 \times 7$$

$$= 22 \times 4$$

$$= 88 \ cm^3$$

Try this 8

Find the volume of the cylinder with radius 3 cm and height 6.3 cm

Exercise 9.3

1. Calculate the curved surface area of each of the following cylinders if:

(a) radius = 7 cm, height = 3 cm

(b) radius = 14 m,   height 5 m

(c) diameter = 4.2 cm, height 5 cm

 (d) diameter = 5 m, height = 4 m

2. Calculate the curved surface area of a cylinder when the diameter and height are 14 cm and 8 cm respectively.

3. Calculate the surface area of each of the following cylinders if:

(a) radius = 2.8 cm, height = 15 cm

(b) radius = 10 m, height = 4 cm

(c) diameter = 10 cm, height = 7 cm

(d) diameter = 28 m, height = 13 m

4. What area of sheet metal is needed to make a closed cylindrical can of radius 2 cm and height 5 cm?

5. What is the area of fencing needed to put a fence 2.5 m high round a circular compound of radius 100 m?

6. What is the area of a label 2 cm wide that goes completely round a can of radius 2.8 cm?

7. What will it cost to paint the curved surface of a cylindrical container that is 30 cm high and 14 cm in diameter at GH¢ 2.50 per square centimetre?

8. A water tank in the shape of a cylinder has an open top. The height of the tank is 28 cm and the diameter is 12 cm. How much will it cost to paint the tank at GH¢ 3.50 per square centimetre?

9. Using $\pi = 3.142$, calculate the volume of each of the following cylinders if:

(a) height = 6 cm, radius = 4 cm   (b) height = 2 m, diameter = 3 m

(c) height = 30 mm, diameter = 24 mm

(d) height = 15 cm, radius = 6 cm

10. The height of a cylinder is 1.8 m and the circumference of the base is 184.8 m. Calculate its volume.

11. The base radius of a cylinder is 2 cm and its total surface area is 40 cm^2. Calculate its height and its volume.

12. Find the volume of a cylindrical tank that has a diameter of 20 cm and a height of 18 cm. (Use $\pi = 3.142$)

13. A cylindrical tank is 8 cm high and 28 cm in diameter. Calculate correct to one decimal place the volume in litres of water it can hold.

14. A cylindrical petrol storage tank is 3 m in diameter and 1.2 m high. How much will it cost to fill the tank if a litre of petrol cost GH¢ 2.50. [Take $\pi = 3.142$]

15. One litre of liquid is poured into a cylindrical container 16 cm

in diameter. Calculate the depth of liquid in the container. Correct your answer to one decimal place.

9.4 Cones

A cone is a solid that has a circular base and a single vertex

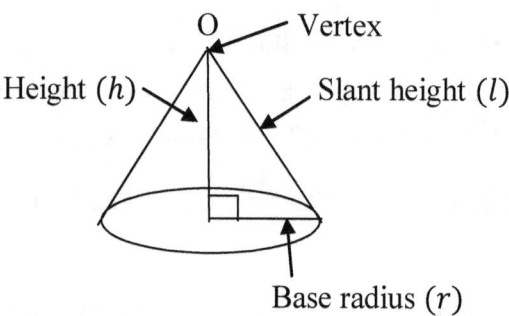

Figure 9.8

Figure 9.8 shows a right circular cone. It has a circular base and a single vertex at O. The line segment from the vertex perpendicular to the base is called the height, h of the cone. The line segment joining the vertex of a cone to the edge of the circular base is called the slant height, ℓ.

The Curved Surface Area

The net of a cone is a sector of a circle. The radius of the sector is the same as the length, ℓ, of the slant height of the cone as shown in Figure 9.9.

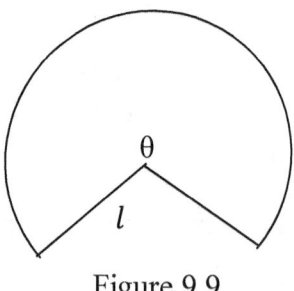

Figure 9.9

The length of the arc of the sector with an angle θ shown in Figure 9.9 is equal to the circumference of the base of the original cone i.e. $2\pi r$. Therefore

$$\frac{\theta}{360} \times 2\pi l = 2\pi r$$

$$\frac{\theta}{360} = \frac{r}{l}$$

The surface area, A of the cone is the same as the area of the sector. Therefore

$$A = \frac{\theta}{360} \times \pi l^2$$

$$= \frac{r}{l} \times \pi l^2$$

$$= \pi r l$$

The slant height, ℓ can be calculated from the relation $l^2 = r^2 + h^2$

Example

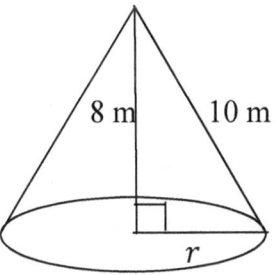

The diagram above shows a cone whose height is 8 m and slant height is 10 m. Find its curved surface area.

$l^2 = r^2 + h^2$

$10^2 = r^2 + 8^2$

$r^2 = 36$

$$r = 6\,m$$

Now the area of the cone is

Curve surface area of cone $= \pi r l$

$$= \frac{22}{7} \times 6 \times 10$$

$$= 188.6\,m^2$$

Try this 9

Calculate the curved surface area of a cone when the height is 12 cm and the radius is 5 cm.

The Surface Area

If the cone is solid, then the surface area will be the curved surface area plus the area of the base. Therefore, the surface area, A, of a cone is

$$A = \pi r l + \pi r^2$$

$$= \pi r (l + r)$$

Example

The base radius of a cone is 8 cm and its height is 15 cm. Calculate the surface area of the cone. [Take $\pi = 3.142$]

$$l^2 = r^2 + h^2$$

$$l^2 = 8^2 + 15^2$$

$$= 64 + 225$$

$$l = 17\,cm$$

Surface area $= \pi r (l + r)$

$$= 3.142 \times 8 \times (17 + 8)$$

$$= 628.4 \ cm^2$$

Try this 10

The slant height of a cone is 13 cm and the height is 12 cm. Find the total surface area of the cone. [Take $\pi = 3.142$]

Volume of a Cone

The volume, V, of a cone with height h and base radius r is given by the formula

$$V = \frac{1}{3}\pi r^2 h$$

Example

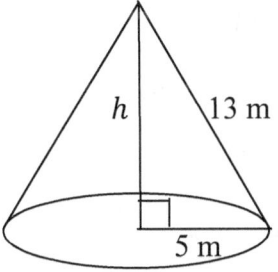

The diagram above shows a cone with radius 5 m and slant height 13 m. Calculate the volume of the cone. [Take $\pi = 3.142$]

$$l^2 = r^2 + h^2$$

$$13^2 = 5^2 + h^2$$

$$h^2 = 144$$

$$h = 12\ m$$

$$Volume\ of\ cone = \frac{1}{3}\pi r^2 h$$

$$= \frac{1}{3} \times 3.142 \times 5^2 \times 12$$

$$= 314.2\ m^2$$

Try this 11

Find the volume of a cone with height 15 cm and slant height 17 cm.

Exercise 9.4

1. Calculate the curved surface area of each of the following cones if:

(a) slant height = 8 cm, base radius = 6 cm

(b) slant height =13 cm, height = 12 cm

(c) slant height = 7 cm, perimeter of the base =15 cm

2. Calculate the total surface area of each of the following cones if:

(a) slant height = 4 cm, base radius = 3 cm

(b) slant height = 26 cm, height = 24 cm

 (c) height = 8 cm, base diameter = 12 cm

3. A sector of a circle of angle 120^0 is folded to form a cone. If the radius of the circle is 21 cm calculate the outer surface area and height of the cone formed.

4. The total surface area of a cone is 1020 square metres and the radius is 14.5 metres. Calculate the slant height.

5. The curved surface area of a cone is 2310 square metres and the radius is 21 metres. Calculate the height of the cone.

6. Calculate the volume of each of the following cones if:

(a) height = 4 cm, base area = 15 cm^2

(b) height = 8 cm, base radius = 3 cm

(c) height = 3 cm, slant height = 5 cm

(d) height = 5 cm, perimeter of base = 8 cm

7. A cone is formed from a sector containing an angle of 216^0 of a circle radius 2 cm. Find the height and volume of the cone formed.

8. The height of a cone is 6 cm. Calculate the radius of the base when the volume is 77 cm^3.

9. The base of a conical tent is 3.6 m in diameter and the height is 1.5 m. Find its volume and the area of the canvas used for making it.

10. The area of the base of a cone is 78.55 cm^2. Calculate the curved surface area when the volume is 314.2 cm^3. [Take $\pi = 3.142$]

11. Find the volume of a cone if the area of the curved surface is 47.1 cm^2 and the base radius is 3 cm. [Take $\pi = 3.142$]

12. The base diameter of a cone is 14 cm and its slant height is 25 cm. What is the height of the cone? Using $\pi = 3\frac{1}{7}$ calculate to three significant figures its curved surface area and volume.

13. A rectangular solid block 12 cm wide, 15 cm long and 8 cm high is melted and recast into a solid right circular cone of base radius 12 cm. Calculate the height of the cone.

14. A solid right prism has a triangular base with sides 8 cm, 15 cm and 17 cm. The height of the prism is 12 cm. If the prism is melted and recast into a solid right circular cone of height 16 cm, calculate the radius of the base of the cone.

15. A right solid cone radius 10.5 cm and height 15 cm is melted and recast into a solid rectangular block. If the base of the block measures 12 cm by 15 cm calculate the height of the block.

9.5 Pyramids

A pyramid is a solid that has a base which can be any polygon and triangular faces that meet at one point called the vertex. The triangular sides are called the lateral faces.

A pyramid is named after its base. A pyramid on a rectangular base is called a rectangular pyramid and a pyramid on a triangular base is called a triangular pyramid or a tetrahedron.

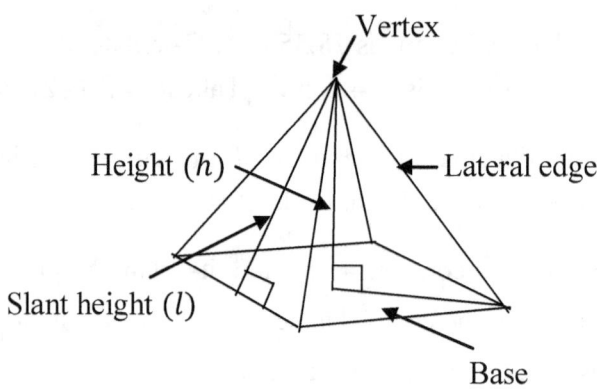

Figure 9.10

Figure 9.10 shows a pyramid on a rectangular base. The line segment joining the vertex to a corner of the base is called the lateral edge (or slant edge) and a line segment from the vertex perpendicular to the base is called the height. The height of each lateral face is called the slant height, l of the pyramid. The lateral faces are congruent isosceles triangles.

Regular Pyramid

A pyramid is called a regular pyramid if its base is a regular polygon and the lateral edges are all equal. The height of a regular pyramid is the line segment joining the vertex to the centre of the base.

Calculating Lengths and Angles

To calculate lengths and angles of pyramids, first identify the appropriate right-angled triangles and then use the Pythagoras' Theorem or the appropriate trigonometric ratio to find the required lengths or angles.

Examples

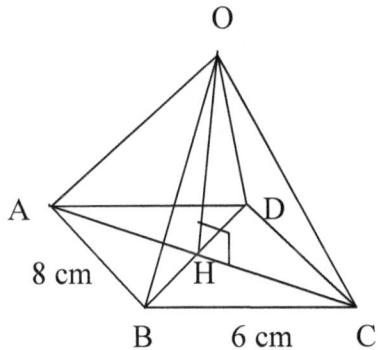

The diagram above shows a rectangular pyramid in which AB = 8 cm, BC = 6 cm and OH = 12 cm. Calculate the length of OA.

Using triangle ABC and Pythagoras' Theorem we have

$$AC^2 = 8^2 + 6^2$$

$$= 100$$

$$AC = 10 \ cm$$

Since the diagonal bisect each other $AH = 5 \ cm$

You can obtain the same result as follows:

The line segment from the point H to the mid-point of AB is perpendicular to AB and is one-half the length of BC. Similarly, the line segment from H to BC is one-half the length of AB.

$$AH^2 = 4^2 + 3^2$$

$$= 25$$

$$AH = 5\ cm$$

Using triangle OHA and the Pythagoras' Theorem we have

$$OA^2 = 5^2 + 12^2$$

$$= 169$$

$$OA = 13\ cm$$

Try this 12

A right pyramid OABCD with vertex O stands on a rectangular base ABCD. AB = 1.2 m, BC = 1.6 m and OA = 2.6 m. Find the height of the pyramid.

A right pyramid OABCD with vertex O stands on a rectangular base ABCD. AB = 12 cm, BC = 16 cm and OH = 13 cm. Calculate $\angle OAH$.

First draw the diagram of the pyramid as shown below.

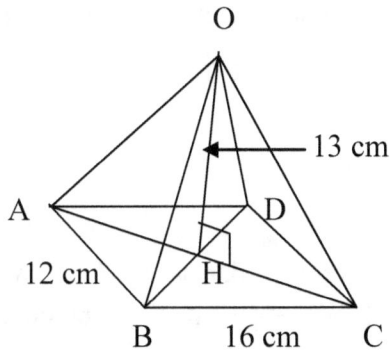

Using triangle ABC and the Pythagoras' theorem we have

$$AC^2 = 12^2 + 16^2$$

$$= 144 + 256$$

$$= 400$$

$$AC = 20\ cm$$

AH is half the length of AC, so

$$AH = 10\ cm$$

Using triangle OAH and the tangent ratio we have

$$\tan \angle OAH = \frac{OH}{AH}$$

$$= \frac{13}{10}$$

$$= 1.3$$

Hence, $\angle OAH = 52.4°$

Try this 13

A right pyramid OABCD with vertex O stands on a rectangular base ABCD. AB = 12 cm, BC = 16 cm and OA = 13 cm. Calculate $\angle OAH$.

Exercise 9.5(a)

1. A right pyramid has a rectangular base WXYZ with vertex O and height OH. Given that WX = 40 cm, XY = 60 cm and OH = 60 cm, find OZ.

2. A right pyramid VABCD with vertex V stands on a square base ABCD of side 10 m. If the length of its slant edges is 15 m, calculate; correct to the nearest unit the height of the pyramid.

3. A pyramid OABCD has a square base ABCD of side 8 m and each of its lateral edges is of length 9 cm. Calculate the height of the pyramid.

4. O is the vertex of a right pyramid on a square base ABCD. Calculate, correct to one decimal place the length of the side of the base of the pyramid when the height is 5 m and its lateral edges are of length 13 m.

5. O is the vertex of a right rectangular pyramid OABCD. AB = 12 cm, BC = 16 cm and OH = 15 cm. Calculate $\angle OAH$

6. O is the vertex of a right rectangular pyramid OABCD. AB = 16m, and BC = 12 m. If the height of the pyramid is 20 m, calculate $\angle OBH$

7. O is the vertex of a right rectangular pyramid OABCD. If AB = 12 cm, BC = 16 cm and OH = 15 cm, calculate $\angle OAB$.

8. A right pyramid OABCD with vertex O stands on a rectangular base ABCD. AB = 10 cm, BC = 24 cm and the height OH = 13 cm. Calculate ∠*OCD*.

9. The lateral edge of a pyramid which stands on a square base is 16 cm. Calculate the height of the pyramid if the angle between a sloping edge and the base is 50⁰

10. The lateral edge of a pyramid which stands on a square base is 18 cm. Calculate the length of the base if the angle between a lateral edge and the base is 60⁰.

Angles between a lateral face and the base

The angle between two planes is the measure of the angle between two lines drawn, one in each plane, perpendicular to the line of intersection of the two planes as shown in Figure 9.11.

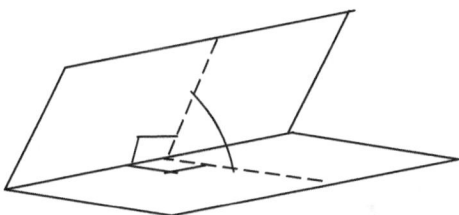

Figure 9.11

The lateral faces of a right pyramid are isosceles triangles. Recall that a line segment from a vertex of an isosceles triangle to the mid- point of the base is perpendicular to the base.

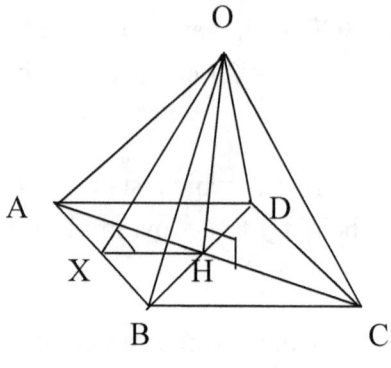

Figure 9.12

In Figure 9.12, X is the mid-point of AB. OX is perpendicular to AB. Because HX is perpendicular to AB we take ∠*OXH* as the measure of the angle between the face OAB and the base. In general, the angle between a lateral face and the base is the size of the angle between the slant height and the line segment from the point of intersection of the diagonals on the base to the mid-point of the edge of the base.

Example

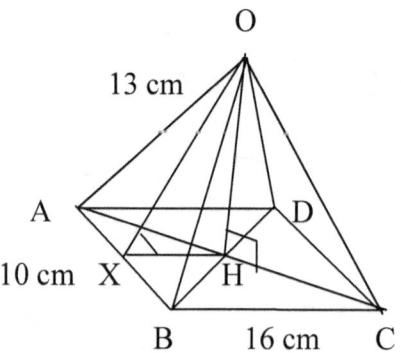

In the diagram above OABCD is a right rectangular pyramid. AB = 10 cm, BC = 16 cm and OA = 13 cm. Calculate the angle between the face OAB and the base.

Using triangle OAX and the Pythagoras' Theorem we have

$$OX^2 + AX^2 = OA^2$$

$$OX^2 + 5^2 = 13^2$$

$$OX^2 = 144$$

$$OX = 12 \ cm$$

Using triangle OHX we have

$$\cos \angle OXH = \frac{XH}{OX}$$

$$= \frac{8}{12}$$

$$= 0.6667$$

Hence, $\angle OXH = 48.2°$

The angle between the face OAB and the base is 48.2^0.

Try this 14

OABCD is a right rectangular pyramid with vertex at O and base ABCD. AB = 6 cm, BC = 10 cm and OH = 12 cm. Calculate the angles between the lateral faces and the base.

Exercise 9.5(b)

1. O is the vertex of a right rectangular pyramid OABCD. AB = 8 cm, BC = 12 cm and OH = 15 cm. Calculate the angle between the lateral faces and the base.

2. The lateral edge of a right rectangular pyramid OABCD with vertex at O is 16 cm. Calculate the angle between the lateral faces and the base if AB = 20 cm and BC = 18 cm.

3. The lateral edge of a right rectangular pyramid OABCD with vertex O is 20 cm. Calculate the angle between the lateral faces and the base if AB = 10 cm and BC = 18 cm.

4. A right pyramid OABCD with vertex O has a square base of length 8 cm, and a lateral edge of 12 cm. Calculate the angle between the lateral face and the base.

The Surface Area of a Pyramid

The total surface area of a regular pyramid is the sum of the lateral area and the area of the base.

Example

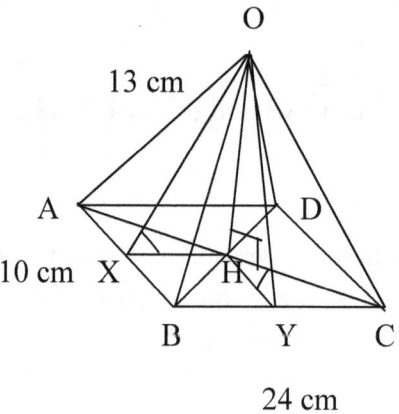

The diagram is a right rectangular pyramid. AB = 10 m, BC = 24 m and OA = 13 m. Calculate the total surface area

Using triangle OAX and the Pythagoras' Theorem

$$OX^2 + AX^2 = OA^2$$

$$OX^2 + 5^2 = 13^2$$

$$OX^2 = 144$$

$OX = 12\ m$

Now use the formula $A = \frac{1}{2}bh$ to find the area of triangle OAB.

$Area\ of \Delta OAB = \frac{1}{2} \times 10 \times 12 = 60\ m^2$

Next calculate the area of face OBC

$OY^2 + BY^2 = OB^2$

$OY^2 + 12^2 = 13^2$

$OY^2 = 25$

$OY = 5\ m$

$Area\ of\ \Delta OBC = \frac{1}{2} \times 24 \times 5$

$= 60\ m^2$

Finally, find the area of the base

$Area\ of\ ABCD = 10 \times 24 = 240\ m^2$

Notice that, the lateral faces OAB and ODC have equal area, and the lateral faces OBC and OAD have equal area.

Hence, $the\ total\ surface\ area = 240 + 2(60 + 60) = 480\ m^2$

Try this 15

A right rectangular pyramid OABCD with vertex O has height 15 cm. AB = 12 cm and BC = 8 cm. Calculate, correct to one decimal place, the total surface area of the pyramid.

The Volume of a Pyramid

The volume of a pyramid is one third of the base area × the perpendicular height. For a rectangular pyramid with base area, lw the volume, V, is

$$V = \frac{1}{3}lwh$$

Similarly for a square pyramid with base length, l the volume, V, is

$$V = \frac{1}{3}l^2h$$

Example

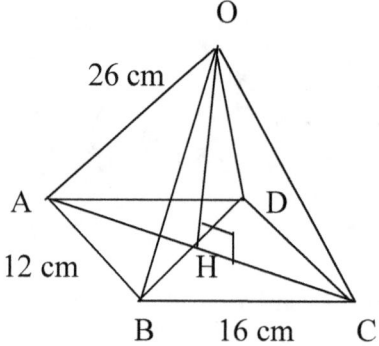

The diagram is a right rectangular pyramid. AB = 12 cm, BC = 16 cm and OA = 26 cm. Calculate the volume of the pyramid.

Using triangle ABC and the Pythagoras' Theorem we have

$$AC^2 = AB^2 + BC^2$$

$$= 12^2 + 16^2$$

$$= 400$$

$$AC = 20\ cm$$

Since AH is one- half of AC

$AH = 10\ cm$

Using triangle OAH and Pythagoras' Theorem

$OH^2 + AH^2 = OA^2$

$OH^2 + 10^2 = 26^2$

$OH^2 = 676 - 100$

$= 576$

$OH = 24\ cm$

The height of the pyramid is 24 cm

$Area\ of\ base = 12 \times 16$

$= 192\ \text{cm}^2$

$Volume\ of\ pyramid = \dfrac{1}{3} Ah$

$= \dfrac{1}{3} \times 192 \times 24$

$= 1536\ cm^3$

Try this 16

O is the vertex of a right rectangular pyramid OABCD. AB = 5 cm, BC = 6 cm and OA = 8 cm. Calculate the volume of the pyramid

Exercise 9.5(c)

1. A right rectangular pyramid OABCD with vertex O has lateral edge 12 cm. AB = 8 cm and BC = 6 cm. Calculate the total surface area of the pyramid.

2. O is the vertex of a right rectangular pyramid OABCD. AB = 10 cm, BC = 12 cm and OH = 15 cm. Calculate the total surface area.

3. A right square pyramid has a lateral edge 17 cm and height 15 cm. Calculate the total surface area.

4. A right pyramid has a square base of side 18 m and height 24 m. Calculate the surface area of the pyramid.

5. The base area of a right square pyramid is 36 cm^2. If the height is 15 cm, calculate the total surface area.

6. O is the vertex of a right rectangular pyramid AB = 12 cm, BC = 5 cm and OA = 13 cm. Calculate the volume of the pyramid.

7. The base ABCD of a right pyramid, vertex O, is a square of side 4 cm, and the length of a lateral edge is 5 cm. Calculate the volume of the pyramid.

8. A right square pyramid has height 6 cm, and the slant height is 9 cm. Calculate the volume of the pyramid.

9. A right pyramid has a square base of side 8 m, and the slant height is 15 m, calculate the volume of the pyramid.

10. A right pyramid has a square base of side 10 m. The faces slope at an angle of 60^0. Calculate the volume of the pyramid.

11. V is the vertex of a right pyramid on a square base ABCD of side 8 m. If the volume of the pyramid is 128 m^3, calculate its height and the angle, to the nearest degree between the base ABCD and a lateral edge of the pyramid.

12. A solid right pyramid with vertex O has a square base ABCD. The volume of the pyramid is 600 m^3 and its height is 8 m. Calculate the length of a side of the base and the angle between a lateral face and the base ABCD, correct to the nearest degree.

9.6 Spheres

A sphere is a solid figure bounded by a curve surface which is such that all points on the surface are a given distance, r, called the radius from a fixed point within the solid called the centre of the sphere.

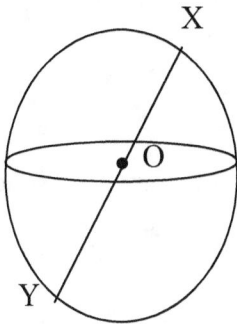

Figure 9.13

Figure 9.13 shows a sphere centre O. If X and Y are points on the surface of the sphere, then $\overline{OX}$ is a radius and $\overline{XY}$ is a diameter of the sphere. For a sphere of radius r the area A is

$$A = 4\pi r^2$$

and the volume, V is

$$V = \frac{4}{3}\pi r^3$$

Example

Find the surface area and volume of a sphere radius 2.1 cm.

$Surface\ area = 4\pi r^2$

$$= 4 \times \frac{22}{7} \times 2.1^2$$

$$= 55.4 \ cm^2$$

$$Volume = \frac{4}{3}\pi r^3$$

$$= \frac{4}{3} \times \frac{22}{7} \times 2.1^3$$

$$= 38.8 \ cm^3$$

Try this 17

Calculate the surface area and volume of a sphere radius 1.4 m

Hemisphere

A hemisphere is obtained by cutting a sphere into two equal halves by a plane passing through its centre.

Therefore, the curved surface area, A of a hemisphere is

$$A = 2\pi r^2$$

and the volume, V is

$$V = \frac{2}{3}\pi r^3$$

The total surface area, A of a solid hemisphere is

$$A = 2\pi r^2 + \pi r^2 = 3\pi r^2$$

Example

Find the volume and total surface area of a hemisphere radius 1.4 m

Volume of hemisphere $= \frac{2}{3}\pi r^3$

$$= \frac{2}{3} \times \frac{22}{7} \times 1.4^3$$

$$= 5.75 \ m^3$$

Total surface area $= 3\pi r^2$

$$= 3 \times \frac{22}{7} \times 1.4^2$$

$$= 18.48 \ m^2$$

Try this 18

Find the volume and the surface area of a hemisphere radius 2.8 m

Exercise 9.6

1. Calculate the surface area and volume of each of the following spheres when the radius is:

(a) 8 cm   (b) 4 m

2. Calculate the surface area and volume of each of the following spheres when the diameter is:

(c) 12 cm   (d) 5.6 m

3. Calculate the curve surface area, surface area and volume of each of the following hemispheres when the radius is:

(a) 3 m   (b 9 cm

4. Calculate the surface area, total surface area and volume of each of the hemispheres when the diameter is:

(c) 14 cm   (d) 5 m

5. Calculate the radius of a sphere when its volume is 4256 cm^3

6. Calculate the radius of a sphere when its surface area is 28.3 cm^2

7. Calculate the volume of a sphere when its surface area is 74.5 cm^2

8. Calculate the surface area of a sphere which volume is 523.7 cm^3

9. The volume of a sphere is 905 cm^3, calculate its surface area

10. Calculate the diameter of a sphere of volume 4851 cm^3

11. The circumference of a sphere is 9 cm. Calculate the surface area of the sphere.

12. The area of a great circle of a hemisphere is 227.0 m^2. Calculate the surface area of the hemisphere.

13. Four spheres of metals each of radius 3 cm are melted and formed into a single cone. If the height of the cone is 15 cm, calculate correct to one decimal place, the radius of the base.

14. A cylindrical measuring glass contains water up to the mark 12.6 cubic centimetres. When a solid metal sphere is dropped into it, the water rises to the mark 14.7 cubic centimetres. Calculate the diameter of the sphere

9.7 Frustum

A frustum is a part of a solid that remains after the top portion has been cut off by a plane parallel to the base.

Figure 9.14 shows the frustum of a cone and Figure 9.15 shows the frustum of a pyramid.

Height   Height

Figure 9.14   Figure 9.15

The height of a frustum is the perpendicular distance between the bases, as shown in Figure 9.14 and Figure 9.15. When a cone (or a regular pyramid) is cut by a plane parallel to the base, a cone (or a regular pyramid) similar to the original is formed.

Calculating the Surface Area of a Frustum

To find the area of a frustum we subtract the area of the portion cut off from the area of the original solid.

Example

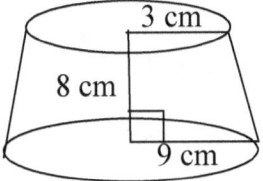

The diagram above shows a frustum of a cone. Calculate its curved surface area to the nearest unit.

The original cone is illustrated below. The top portion of the cone is similar to the original cone.

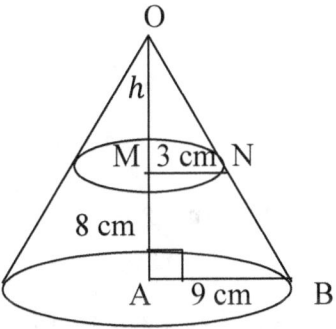

Notice that $\triangle OMN$ and $\triangle OAB$ are similar triangles. Hence

$$\frac{h}{h+8} = \frac{3}{9}$$

$$h = 4 \text{ cm}$$

Find the slant height of the top portion of the cone

$$ON^2 = 4^2 + 3^2$$

$$= 25$$

$$ON = 5 \text{ cm}$$

Find the curved surface area of the top portion

Curve surface area of top portion $= \pi r l$

$$= 3.142(3)(5)$$

$$= 47.13 \text{ cm}^2$$

Next find the area of the original cone

$$\frac{ON}{OB} = \frac{3}{9}$$

so $OB = 3(5)$

$$= 15 \text{ cm}$$

Curved surface area of original cone $= 3.142(9)(15)$

$$= 424.17 \text{ cm}^2$$

Finally, subtract the area of the top portion from the area of the original cone

Curved surface area of frustum $= 424.17 - 47.13$

$$= 377.04 \text{ cm}^2$$

The curved surface area of the frustum is 377 cm^2

The curved surface area of the original cone (or the top portion) can also be obtained as follows:

Recall that if the ratio of the corresponding sides of two similar solids is $a : b$ then the ratio of their area is $a^2 : b^2$ and the ratio of their volume is $a^3 : b^3$.

So $\dfrac{curved\ surface\ area\ of\ top\ portion}{curved\ surface\ area\ of\ original\ cone} = \left(\dfrac{3}{9}\right)^2$

Curved surface area of original cone = 9 × curved surface area of top portion

$$= 9 \times 47.13$$

$$= 424.17 \text{ cm}^2$$

Note that the surface area of the frustum of the cone is the area of the curved surface plus the area of the bases

Calculating the Volume of Frustum

To find the volume of a frustum we subtract the volume of the portion cut off from the volume of the original solid.

Try this 19

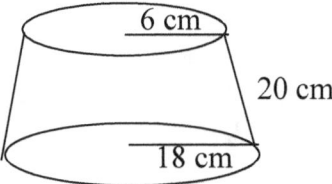

Calculate the volume of the frustum of the cone shown above.

Exercise 9.71.

Calculate the curved surface area and volume of the original cone of each frustum.

(a)   (b)

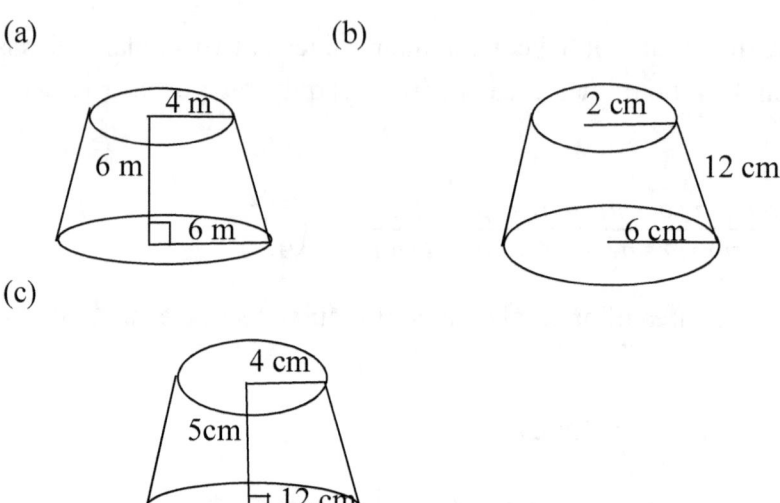

(c)

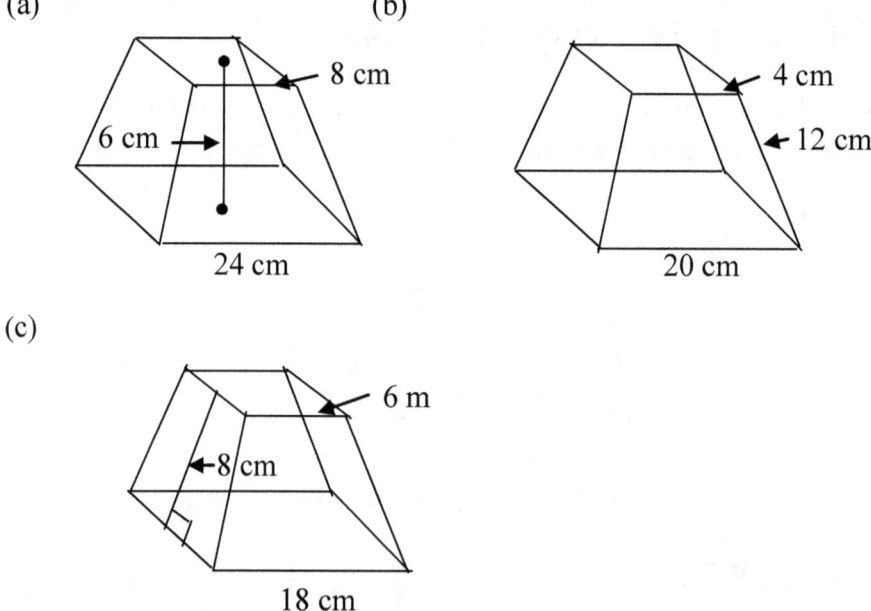

2. Calculate the lateral area and volume of the original regular pyramid of each frustum

3. Calculate the curved surface area and volume of the top portion of the original cone of each frustum

(a)   (b)

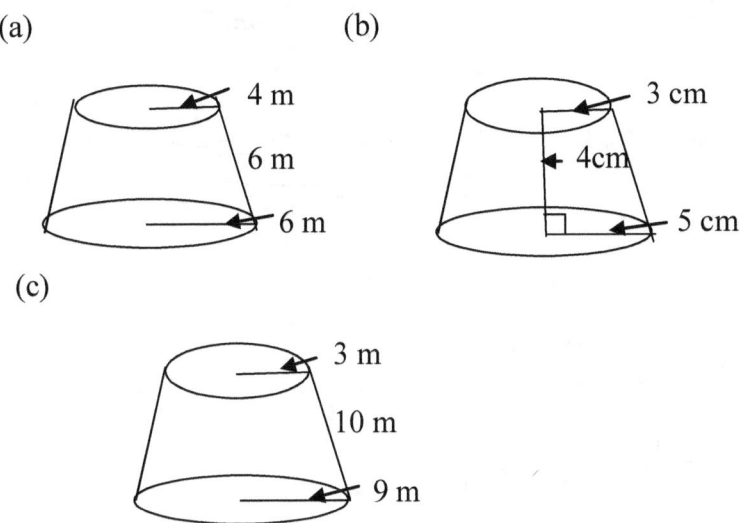

(c)

4. Calculate the lateral area and volume of the top portion of the original regular pyramid of each frustum

(a)   (b)

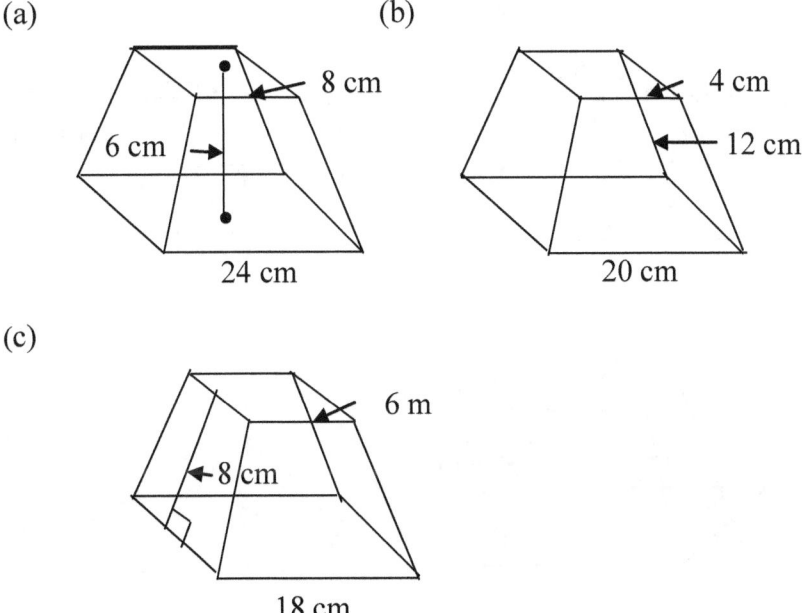

(c)

5. Calculate surface area and volume of the frustum of each cone

(a)   (b)

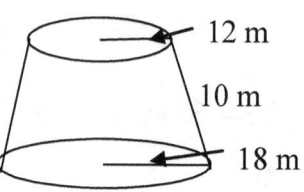

12 m

10 m

18 m

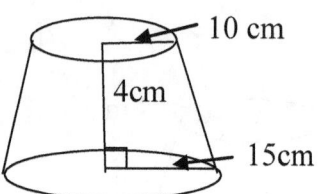

10 cm

4cm

15cm

(c)

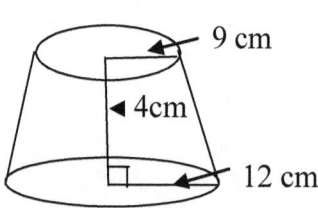

9 cm

4cm

12 cm

6. Calculate the lateral area and volume of the frustum of each regular pyramid

(a)   (b)

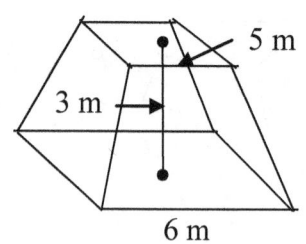

5 m

3 m

6 m

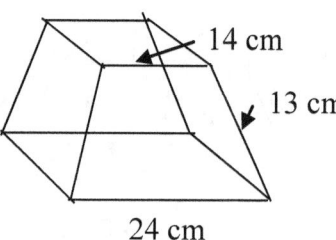

14 cm

13 cm

24 cm

(c)

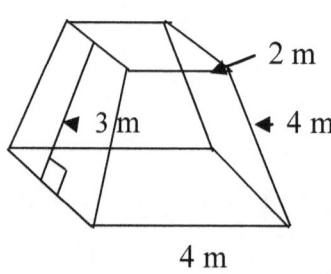

2 m

3 m

4 m

4 m

9.8 Composite Solids

A composite solid is a solid formed by two or more solids.

The Surface Area of a Composite Solid

To calculate the area of a composite solid, calculate the area of each of the individual solids, and then add their areas together to give the area of the composite solid.

Example

Calculate the surface area, correct to one decimal place, of the solid shown below. [Take $\pi = 3.142$].

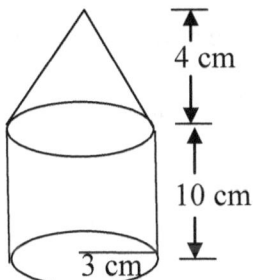

The solid is a composite of a cone and a cylinder. The base of the solid is the base of the cylinder. First, calculate the surface area of the cone. Notice, that the base radius of the cone is the same as the base radius of the cylinder.

Here $r = 3$ cm and $h = 4$ cm. If l is the slant edge of the cone then

$l^2 = r^2 + h^2$

$l^2 = 3^2 + 4^2$

$\quad = 25$

$l = 5 \; cm$

Area of cone $= \pi r l$

$$= 3.142 \times 3 \times 5$$

$$= 141.39 \; cm^2$$

Next calculate the surface area of the cylinder

Surface area of cylinder $= 2\pi r h + \pi r^2$

$$= \pi r(2h + r)$$

$$= 3.142 \times 3 \times (20 + 3)$$

$$= 216.8 \; cm^2$$

Finally add together the area of the cone and the cylinder

Surface area of solid $= 141.39 + 216.8$

$$= 358.16$$

$$= 358.2 \; cm^2$$

The surface area of the solid is 358.2 cm²

Try this 20

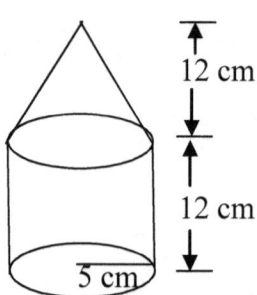

The solid is made up of a cone and a cylinder. Find the surface area, correct to one decimal place of the solid above. [Take $\pi = 3.142$].

The Volume of a Composite Solid

To calculate the volume of a composite solid, calculate the volume of each individual solid, and then add their volumes together to give the volume of the composite solid.

Example

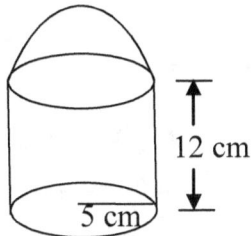

The solid above is made up of a hemisphere and a cylinder. Find the volume of the solid, correct to the nearest unit. [Take $\pi = 3.142$].

First, calculate the volume of the hemisphere.

$$\text{Volume of hemisphere} = \frac{2}{3}\pi r^3$$

$$= \frac{2}{3} \times 3.142 \times 5^3$$

$$= 261.8 \ cm^3$$

Next, calculate the volume of the cylinder

$$\text{Volume of cylinder} = \pi r^2 h$$

$$= 3.142 \times 5^2 \times 12$$

$$= 942.6 \ cm^2$$

Volume of solid $= 261.8 + 942.6$

$$= 1204.4$$

$$= 1204 \ cm^3$$

The volume of the solid is 1204 cm^3

Try this 21

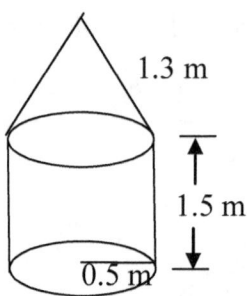

The solid is made up of a cone with slant height 1.3 m   and a cylinder radius 0.5 m and height 1.5 m. Find the volume of the solid correct to three significant figures. [Take $\pi = 3.142$].

Exercise 9.8

1. A solid consists of a hemisphere, radius 3 cm, joined to a cone of the same base radius and height 4 cm. Calculate its surface area

2. A solid consists of a hemisphere, radius 10 cm, joined to a cone of the same base radius and height 20 m. Calculate its volume,

3. Calculate the surface area of each of the following solids. Round off to the nearest tenth. [$Take \ \pi = 3.142$]

(a)   (b)   (c)

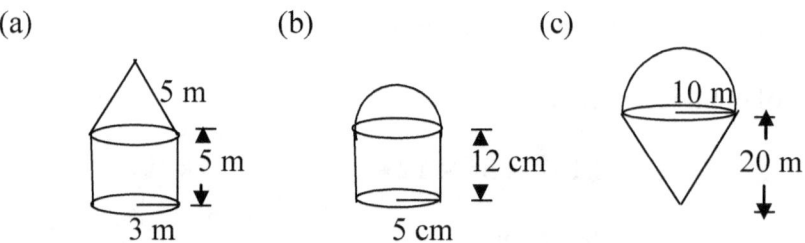

4. Calculate the volume of each of the following solids.   Round off to the nearest tenth. [$Take\ \pi = 3.142$]

(a)   (b)   (c)

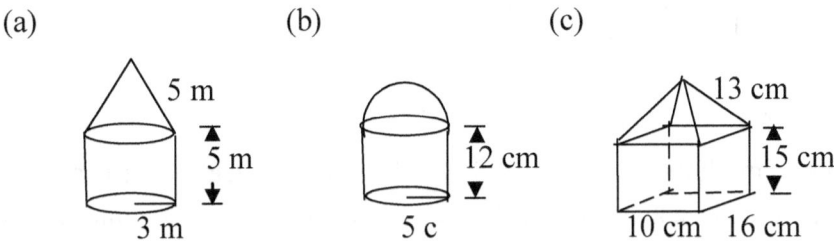

5. A large grain silo is a cylinder with a cone at the bottom. Both the cylinder and cone have a diameter of 8 m.   The height of the cylinder is 28 m and the height of the cone is 3 m. Calculate the volume of grain that may be put in the silo.

6. A solid consists of a hemisphere, radius 8 cm, joined to a cone of the same base – radius and height 6 cm. Calculate the volume and total surface area of the solid.

7. A storage container is in the shape of a hemisphere on top of a cylinder whose base radius is 5 metres and height is 8 metres. The surface of the storage container is to be painted. Calculate the area to be painted.

8. The diagram below shows a petrol tank of a vehicle.

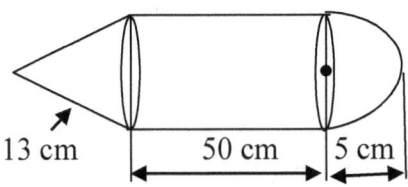

How much will it cost to fill the tank, if a litre of petrol cost GH¢ 8.50?

9.

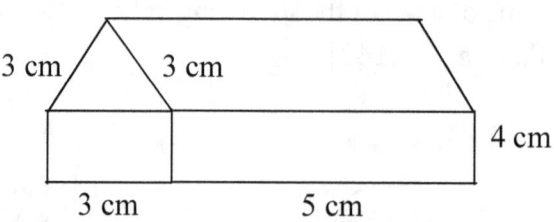

The diagram shows a metal solid which is in the shape of a triangular prism mounted on a cuboid. If the metal is melted and recast into a solid sphere, find the radius of the sphere.

Review exercise 9

1. Calculate the surface area and volume of each of the following right prisms.

(a)   (b)

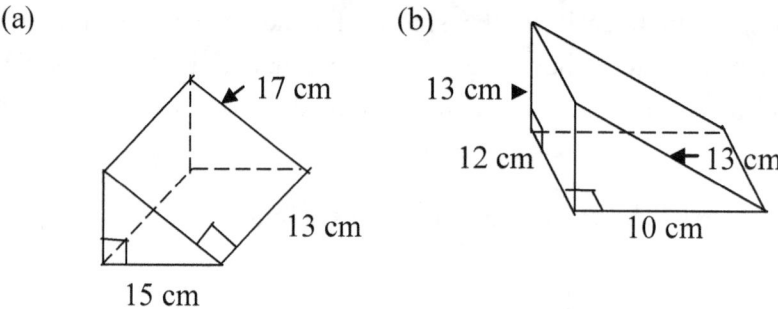

(c)

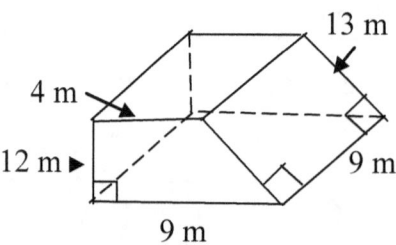

2. The triangular base of a right triangular prism is equilateral with side 6.2 m. If the height of the prism is 20.4 m, find its volume.

3. Calculate the surface area and volume of each of the following cuboids if:

(a) length = 6 cm, width = 3 cm, height = 2.5 cm

(b) length =15 cm, width = 10 cm, height = 5 cm

(c) length = 14.5 cm, width = 7.5 cm, height = 8 cm

(d) length = 3 m, width = 2.5 m, height = 1.5 m

4. A room is 12 m long, 8 m wide and 6 m high. How many boxes can it hold if each box occupies a space of 1.5 m^3?

5. Calculate the length of the edges of a cube when the volume is

(a) 1728 cm^3 (b) 3375 cm^3

6. Calculate the volume of a cube when the surface area is 54 cm^2

7. Calculate the volume of a cube for which a diagonal of one of its faces measures 12 cm.

8. Three cubes each of side 8 cm are joined end to end. Find the surface area of the resulting cuboid.

9. If the volume of a cube is 2197 cm^3, find the surface area and length of the main diagonal of the cube.

10. A wooden box 2.5 m long, 1.2 m wide and 80 cm deep and open at the top is to be made. Determine the cost of wood required for it, if one square metre of wood costs 15 Gp.

11. A tank contains 60,000 m^3 of water. If the tank is 50 m long and 40 m wide, calculate its depth.

Core Mathematics

362

12. Find the length of a rectangular solid of height 6 cm that is twice as long as it is wide, if its volume is the same as that of a cube with a total surface area of 864 cm²

13

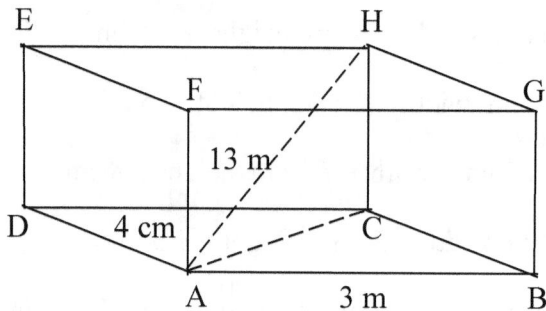

The diagram shows a closed box in the form of a cuboid. AB = 3 m, AD = 4 m and the leading diagonal AH = 13 m. Calculate the total outer surface area and volume of the box.

14. Calculate the curved surface area, total surface area and volume of each of the following cylinders if:

(a) radius = 2.1 cm, height = 10 cm

(b) diameter = 7 m, height = 12 m

(c) radius = 5 m, height = 1.4 m

(d) diameter = 2.8 m, height = 15 m

15. A water storage tank has a cylindrical shape. If it is 2.1 m high and has diameter 1 m, calculate its volume

16. The volume of a right circular cylinder is 3080 cm³ and the radius of base is 7 cm. Calculate the curved surface area of the cylinder.

17. A hollow cylindrical tube, open at both end is made of iron 0.5 cm thick. If the external diameter is 2 cm and the length of the tube is 40 m, calculate the volume of iron used in making the tube

18. A cylindrical bucket of diameter 28 cm and height 12 cm is full of water. The water is emptied into a rectangular tub 66 cm long and 28 cm wide. Calculate the height to which water rises in the tub

19. Calculate the curved surface area, total surface area and volume of each the following right circular cones if:

(a) radius = 5 cm, height = 12 cm

(b) radius = 3 m, slant height = 5 m

(c) diameter = 14 cm , height = 9 cm

(d) diameter = 12 m, slant height = 10 m

20. Calculate the radius of the base of a right circular cone of height 10.5 cm and volume 176 cm^3.

21. A cone has volume 105 m^3 and base radius 7 m. Find its height.

22. A cone has volume 6 cm^3 and height 2 cm. Find the base radius.

23. Calculate the slant height and curved surface area of a cone when the volume is 220 cm^3 and the diameter of the base is 28 cm.

24. Three metal solid cones, each of radius 2 cm and height 5 cm, are melted and formed into a single cube. Calculate to two significant figures, the length of the edge of the cube

25. A cone is formed from a sector containing an angle of 240^0 of a circle radius 14 cm. Calculate correct to three significant figures the outer surface area and height of the cone.[Use $\pi = 3.142$]

26. A sector of angle 240^0 is removed from a thin circular metal sheet of radius 7.5 cm. The remaining portion is then folded, with the straight edges coinciding to form a right circular cone. Calculate its outer surface area and volume.

27. The base of a right pyramid is a rectangle of sides 7 cm and 8 cm. If the height of the pyramid is 9 cm, find the length of a slant edge

28. The base of a right pyramid is a rectangle of sides 12 cm and 9 cm. If the length of its slant edges is 15 cm, find the height of the pyramid.

29. The base of a right pyramid is a square. If the height of the pyramid is 8 cm and the length of its slant edges is 12 cm, find the length of the side of its base.

30. The base of a right pyramid is a rectangle of sides 8 cm and 10 cm. If the height of the pyramid is 12 cm, find the angle between a slant edge and the base.

31. The base of a right pyramid is a rectangle of sides 10 cm and 12 cm. If the length of the slant edge is 15 cm, find the angles between the lateral faces and the base.

32. A right pyramid has a square base of sides 5 cm. If the length of the slant edges is 8 cm, find the angle between the lateral face and the base.

33. The base of a right pyramid is a rectangle of sides 8 cm and 6 cm. If the length of the slant edge is 12 cm, calculate its total surface area.

34. The base of a right pyramid is a rectangle of sides 9 cm and 12 cm. If the height of the pyramid is 15 cm, find its total surface area.

35. The base area of a right square pyramid is 64 cm^2. If the height is 15 cm, calculate the total surface area.

36. The base of a right pyramid is a rectangle of sides 10 cm and 12 cm. If the length of the slant edge is 16 cm, calculate the volume of the pyramid.

37. The base of a right pyramid is a square of side 8 cm. If the length of the slant edge is 10 cm, calculate the volume of the pyramid.

38. The base of a pyramid is a rectangle of sides 7 m and 8 m. If the volume is 112 m³, calculate the height of the pyramid.

39. A pyramid with a square base has height 5 cm and volume 15 cm³. Calculate the side of the base.

40. The area of the smaller base of a truncated cone is 154 cm² and the area of the larger base is 1386 cm². If the height of the truncated cone is 16 cm, calculate its curve surface area and volume.

41. The larger base edge and the smaller base edge of a truncated square pyramid are 50 m and 30 m respectively. If the length of the lateral edge is 12 m, calculate the surface area and volume of the truncated square pyramid.

42.

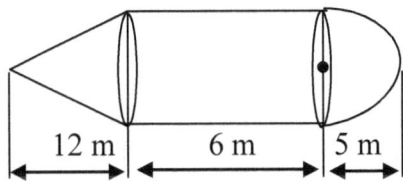

12 m   6 m   5 m

The diagram shows a storage tank made of aluminium sheet. Find the cost of sheet required to make the tank if one square metre of sheet cost GH¢ 6.50.

43. Calculate the surface area and volume of each solid.

(a)   (b)   (c)

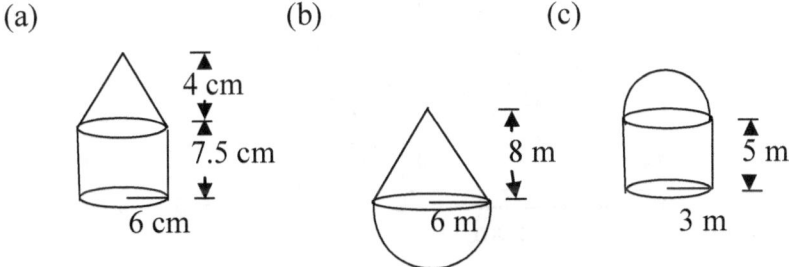

(a) 4 cm 7.5 cm 6 cm

(b) 8 m 6 m

(c) 5 m 3 m

44. Calculate the surface area and volume of each of the following spheres when the radius is:

(a) 10 cm   (b) 6 m   (c) 3.5 m

45. Calculate the radius of each of the following spheres when the volume is:

(a) 12 m^3 (b) 46 cm^3 (c) 523.8 cm^3

46. A sphere has volume 3.44 cm^3. What is its area?

47. A sphere has surface area 66 m^2. What is its volume?

48. 10 dozen lead spheres, each of diameter 3 cm, are melted and recast in the form of a cylinder 12 cm in diameter. Calculate the height of the cylinder.

Chapter Test 9

Take this test as you would take a test in a class. After you are done, check your work against the answers in the back of the book.

1. A prism has height 12 cm and its base is a right- angled triangle with sides 8 cm, 15 cm and 17 cm, find its surface area and volume.

2.

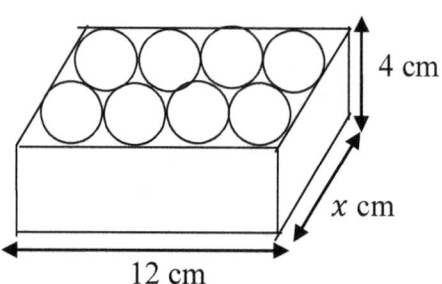

Eight cylindrical can with the same base radius fit exactly inside a box 12 cm long, x cm wide and 4 cm high. Calculate

(a) the value of x

(b) the volume of the box

(c) the volume of box unoccupied

3. Three cubes of metals with edges 4cm, 5 cm and 6 cm respectively are melted and formed into a single cube. If there is no waste in the process, find the edge of the new cube so formed.

4. Water flows through a pipe that has internal diameter of 6 cm. If the water flows at the rate of 1.4 m per second, how many litres of water are delivered in 30 minutes?

5. Find, to 3 significant figures the volume of metal in a hollow pipe 6 cm long, of internal diameter 2 cm, made of metal 0.5 cm thick.

6. A conical vessel of internal radius 1.4 m and height 3.6 cm is full of water. If this water is poured into a cylinder with internal radius 2.1 m, calculate the height to which the water rises in the cylinder.

7. A circular cone is formed from a sector of a circle of arc length 17.6 cm. If the height of the cone is 1.5 cm, calculate the curved surface area.

8. The external diameter of a hollow metal sphere is 12 cm, and its thickness is 2 cm. The sphere is melted and recast in the form of a solid sphere. Calculate the radius of the solid sphere formed.

9. A right square pyramid has slant height of 15 cm. If the lateral area of the pyramid is 540 cm^2, calculate its volume.

10.

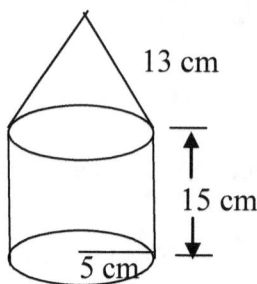

13 cm

15 cm

5 cm

The diagram shows a container made of aluminium sheet. The top part of the container is a cone with slant height 13 cm, and the lower part is a cylinder 15 cm long. The base radius of the cone and cylinder are each 5 cm. Calculate the cost of sheet required to make the container, if one square centimetre of sheet costs GH¢ 3.00.

10

Distances on the Earth

The earth is nearly spherical and its radius is approximately 6,378 km. The Earth's circumference is about 40,000 km.

Figure 10.1 represents the Earth.

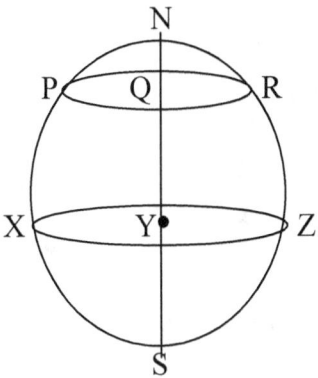

Figure 10.1

The diameter NS is called the axis of the Earth. N is the North Pole, and S the South Pole. The two sections PQR and XYZ are perpendicular to the axis.

Sections perpendicular to the axis such as PQR and XYZ are circles of different sizes. The section through the centre is a great circle. The circles become smaller the further they are from the centre and closer to the poles.

Latitude and Longitude

Any point on the surface of the earth is specified by imaginary lines called latitude and longitude

Lines of Longitude

Lines of longitude, also called the meridians are semi-circles joining the North Pole and the South Pole. The meridian which passes through Greenwich, London called the 'Prime Meridian' is taken as the zero longitude. All other longitudes are specified, as shown in Figure 10.2, by how many degrees east or west they are from the Prime Meridian, using angles from 0° to 180°.

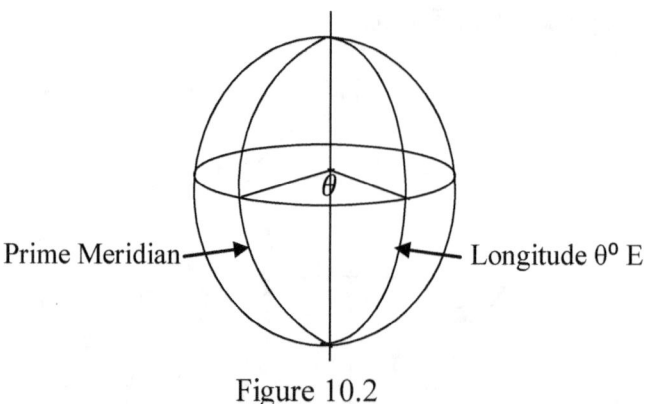

Figure 10.2

Lines of longitude lie on great circles. Recall that a great circle is any circle on the Earth's surface whose radius is the same as that of the earth, and has the same centre as the Earth.

Circles of Latitude

The cross sections of the Earth are a set of circles of different sizes, called the circles of latitude. Circles of latitude are often called parallel of latitudes because they are parallel to each other.

The circle of latitude through the centre of the earth is a great circle, called the equator. The radius of the equator is the same as the radius of the Earth. All other circles of latitudes are small circles.

The equator is equidistant from the North Pole and South Pole. It divides the Earth into two hemispheres, the Northern hemisphere and the Southern hemisphere.

Each circle of latitude is specified, as shown in Figure 10.3, by how many degrees north or south they are from the equator, using angles from 0° to 90°. The North and South Poles are 90° North and 90° South respectively.

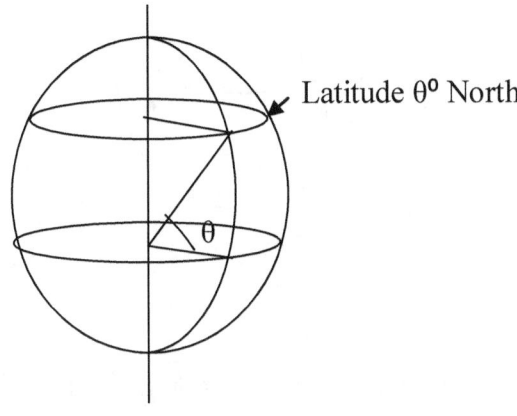

Figure 10.3

Describing Positions on the Surface of the Earth

Any position on the surface of the earth can be described by first specifying which parallel of latitude, and then the longitude the position is on. For instance, a point P on latitude 30° North and longitude 80° East is located at the point of intersection of the line of longitude 80° E and the circle of latitude 30° N, as shown in Figure 10.4. The formal way to specify this position is: (30° N, 80° E).

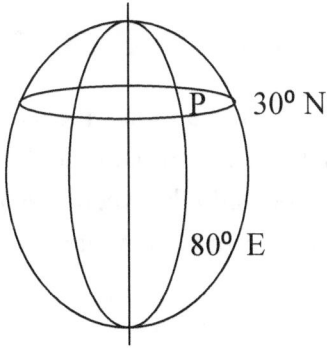

Figure 10.4

Distances on the Earth

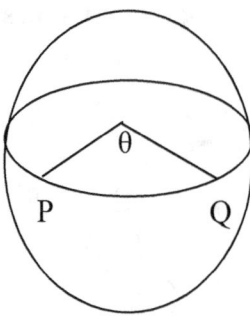

Figure 10.5

Figure 10.5 shows two points P and Q on the same circle of latitude. The two points P and Q separate the circle into two arcs. The length of the minor arc is the shortest distance between the points. The length of the distance, PQ is given by the formula

$$distance\ PQ = \frac{\theta}{360} \times 2\pi r$$

where θ is the angle subtended by the minor arc and r is the radius of the circle of latitude.

10.1 Calculating the radius of a circle of latitude

The only circle of latitude that has a radius the same as the Earth's is the equator. All the other circles of latitude have radii smaller than the Earth's radius. The relationship between the radius of a circle of latitude and the radius of the earth is illustrated below.

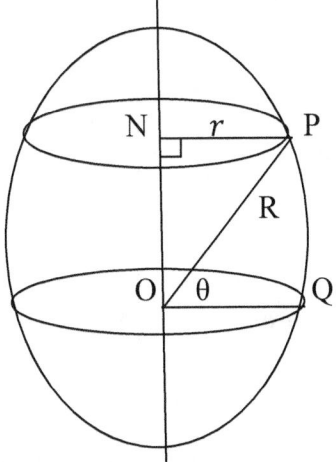

Figure 10.6

Figure 10.6 represents the earth, with O as centre and N as the centre of the θ° N parallel of latitude. The radius r of the circle of latitude θ° N is parallel to the radius R of the equator, and angle OPN is θ° (alternate angles) From triangle ONP

$$\cos \theta^\circ = \frac{r}{R}$$

Multiplying each side of the equation by R, gives

$$r = R\cos \theta^\circ$$

In general, the radius of a small circle is given by the expression

$$r = R\cos \theta^\circ$$

Example

Calculate the radius of circle of latitude 60° N. Take R = 6400 km

The radius, r is given by

$$r = R\cos \theta^\circ$$

Core Mathematics

$$= 6400 \cos 60°$$

$$= 3200$$

The radius of latitude $60°$ is 3200 km.

Try this 1

Calculate the radius of the circle of latitude $49°$

Example 10.1

In these exercises take $\pi = 3.142$ and $R = 6400$ km. Round off answers to the nearest tenth, if necessary.

1. Calculate the radius of the circle of latitude:

(a) $65°$ S   (b) $70°$ N   (c) $15°$ N   (d) $36°$ S

2. Calculate the length of the parallel of latitude:

(a) $48°$ N   (b) $75°$ S   (c) $60°$ N   (d) $55°$ S

3. The radius of a circle of latitude is 1200 kilometre, what is its latitude?

4. The radius of a circle of latitude is 2400 kilometre, what is its latitude?

5. The length of a parallel of latitude is 22,489 kilometre, what is the latitude?

6. The length of a parallel of latitude is 8362 kilometre what is the latitude?

10.2 Finding distances along circles of latitude

Using the arc length formula, you can work out the distance between two points on the same circle of latitude.

Example

P (60° N, 97° W) and Q (60° N, 25° W) are two points on latitude 60° N. Calculate the distance between P and Q, to the nearest unit. [Take $\pi = 3.142$ and $R = 6400$ km]

Figure 10.7, shows a diagram representing the given information. P and Q are located on longitude 97° W and longitude 25° W respectively.

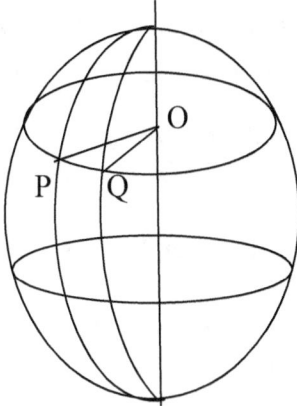

Figure 10.7

Let O be the centre of the 60° N parallel of latitude. The part of the parallel of latitude between P and Q is an arc. The angle at the centre O is the difference in longitude. Hence $\angle POQ = 97° - 25° = 72°$.
Using the arc length formula, we have

$$PQ = \frac{72}{360} \times 2\pi R \cos 60°$$

$$= \frac{72}{360} \times 2 \times 3.142 \times 6400 \times 0.5$$

$$= 4021.76$$

$$= 4022 \text{ km}$$

The distance between P and Q is 4022 km, to the nearest unit.

Try this 2

Two towns A and B, lie on latitude 30^0 S. Their longitudes are 57^0 W and 35^0 E respectively. Calculate the shortest distance between the two towns.

Exercise 10.2

In these exercises take $\pi = 3.142$ and $R = 6400$ km. Round off answers to the nearest tenth, if necessary.

1. Find the difference in latitude between two places P and Q when the latitudes are:

(a) 7^0 N and 68^0 N   (b) 19^0 S and 49^0 N

(c) 25^0 S and 27^0 S   (d) 64^0 N and 76^0 S

2. Two towns P and Q are located on the parallel of latitude 16^0 N, one at longitude 15^0 W and the other at longitude 31^0 E. How far apart are the two towns?

3. Two towns P and Q are located at $(30^0$ N, 70^0 W) and $(30^0$ N, 35^0 W) respectively. Calculate the shortest distance from P to Q along the parallel of latitude 30^0 N.

4. Two towns A and B lie on latitude 48^0 S. Their longitudes are 57^0 E and 87^0 E respectively. Calculate the shortest distance between the two towns.

5. Two cities P and Q, lie on latitude 65^0 N, one at longitude 13^0 W and the other at longitude 12^0 E. Calculate the distance between the two cities, measured along the parallel of latitude.

6. Two cities P and Q are located at $(70^0 \text{ S}, 15^0 \text{ W})$ and $(70^0 \text{ S}, 48^0 \text{ E})$ respectively. Calculate the shortest distance between the two cities.

7. Two towns are located on latitude 50^0 S. The towns are 5,200 kilometres apart. Calculate the difference in their longitude.

8. Two points P and Q are located on the parallel of latitude 25^0 N, P at longitude 30^0 E. If the distance between P and Q is 1500 kilometres, calculate the longitude of Q.

9. From a town located at $(45^0 \text{ N}, 60^0 \text{ W})$, a ship travels 1600 kilometres due east, calculate its new longitude.

10. A ship is at the point 55^0 E on the circle of latitude 36^0 N. If it travels 8500 kilometres due west, calculate its new longitude.

11. Two places are located on the parallel of latitude 35^0 N, one at longitude 45^0 W and the other at longitude 27^0 E. If it takes a plane 12 hours to travel from one town to the other, calculate the speed of the plane, correct to the nearest kilometre per hour.

12. An aircraft flies from a town P located at $(35^0 \text{ N}, 40^0 \text{ W})$ to town Q located at $(35^0 \text{ N}, 30^0 \text{ E})$ at the speed of 540 km h^{-1}. Calculate the time it takes to complete the journey.

10.3 Finding distances along Lines of Longitudes

We now consider distances along lines of longitude. A longitude is a great circle, so its radius is R.

Example

Two points P and Q are located at $(25^0 \text{ N}, 30^0 \text{ E})$ and $(35^0 \text{ S}, 30^0 \text{ E})$ respectively. Calculate the shortest distance between P and Q.

The diagram represent the given information (see Figure 10.8). The point P is at latitude $25°$ N and Q is at latitude $35°$ S. Notice that the part of the line of longitude between P and Q is an arc, which is a part of a circle with the same radius as the radius of the Earth.

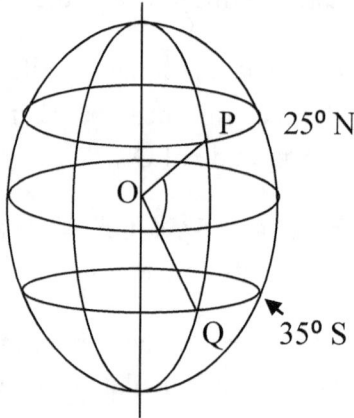

Figure 10.8

The angle between the radii OP and OQ is $(25 + 35)°$, i.e. $60°$. Using the arc length formula, we have

$$PQ = \frac{60}{360} \times 2\pi R$$

$$= \frac{60}{360} \times 2 \times 3.142 \times 6400$$

$$= 6702.93$$

$$= 6703 \text{ km}$$

The distance between P and Q is 6703 km, to the nearest unit

Try this 3

Two towns A and B, lie on longitude 20° E. Their latitudes are 50° N and 30° S, respectively. Calculate the shortest distance between the two towns

Exercise 10.3

In these exercises take $\pi = 3.142$ and $R = 6400$ km. Round off answers to the nearest tenth, if necessary

1. Find the difference in longitude between two places P and Q when the longitudes are:

(a) 32° E and 68° E   (b) 28° E and 65° W

(c) 22° W and 42° E   (d) 75° W and 105° W

2. Two points P and Q are located at (25° N, 30° E) and (35° S, 30° E) respectively. Calculate the shortest distance between P and Q.

3. P and Q are two points on latitude 97° N and 25° N respectively. If both points lie on longitude 60° E calculate their distance apart measured along this longitude.

4. Calculate the distance measured along longitude 55° E from a point on the parallel of latitude 18° N to the South Pole.

5. Calculate the distance from the North Pole to a point P located at (51° N, 35° E) measured along longitude 35° E.

6. Two towns, P and Q lie on longitude 130° W. Their latitudes are 12° N and 33° S respectively. Calculate the shortest distance between the two towns.

7. Two ports, A and B lie on the same longitude. Their latitudes are 45° N and 55° S respectively. Calculate the shortest distance between the two ports.

8. Two towns P and Q are located at (40° N, 65° E) and (30° S, 65° E) respectively. If an aircraft flies from P to Q at the speed of 500 km h⁻¹, calculate the time it takes to complete the journey.

9. Two places are located on longitude 23° W, one at latitude 74° N and the other at latitude 25° N. If it takes a plane 10 hours to travel from one place to the other, calculate the speed of the plane.

10. Two towns are located on longitude 48° E. The towns are 7200 kilometres apart. Calculate the difference in their latitudes.

11. Two towns P and Q lie on the same longitude. Town P lie on latitude 35° S. If Q is due North of P, and 12,500 kilometres away from P, calculate the latitude of Q.

12. Two towns P and Q are located on the same longitude, P at latitude 50° N. If an aeroplane flying due south from P at an average speed of 600 km h⁻¹ reaches Q in 15 hours calculate the latitude of Q.

Review exercise 10

In these exercises take $\pi = 3.142$ and $R = 6400$ km. Round off answers to the nearest tenth, if necessary.

1. Find the difference in latitude between two places A and B when the latitudes are:

(a) 17° N and 58° N   (b) 39° S and 31° N

(c) 52° S and 72° S   (d) 46° N and 67° S

2. Find the difference in longitude between two places P and Q when the longitudes are:

(a) 23° E and 86° E   (b) 22° E and 48° W

(c) 25° W and 95° E   (d) 57° W and 105° W

3. Find the radius of each of the circles of latitude:

 (a) 56° S   (b) 75° N   (c) 25° N   (d) 63° S

4. The length of a parallel of latitude is 32,500 kilometres. What is its latitude?

5. A point P is located at $(40^\circ$ N, 52° W). If a town Q is 1,450 kilometres due East of P, calculate the longitude of Q.

6. Two towns, A and B lie on longitude 38° W. Their latitudes are 51° N and 73° N respectively. Calculate the shortest distance between the two towns.

7. Two towns P and Q lie on the same longitude. Their latitudes are 75° N and 65° S, respectively. Calculate the shortest distance between the two towns

8. Two cities lie on the Equator, one at longitude 56° E and the other at longitude 32° W. Calculate the distance between the two cities, measured along the Equator.

9. Find the distance, measured along a meridian, from any point on parallel of latitude 40° N to the North Pole

10. Two towns are located on longitude 60° W. The towns are 2,500 kilometres apart. Calculate the difference in their latitudes

11. Two places are located on the parallel of latitude 23° N, one at longitude 32° W and the other at longitude 40° E.

(a) How far apart are the two towns, measured along the parallel of latitude 23° N

(b) If it takes an aircraft 15 hours to travel from one town to the other, calculate the speed of the aircraft.

12. Two towns A and B are located at $(32^\circ$ N, 76.2° W) and $(32^\circ$ N, $119.3^\circ)$ respectively. An aircraft flies from A to B at the speed of 1500 km h^{-1}, calculate the time it takes to complete the journey.

13. Two towns P and Q are both on latitude 30° N, and P at longitude 15° W. If an aircraft flying due West from P at an average speed of 450 km h^{-1} reaches Q in 6 hours, calculate the longitude of Q.

14. Two towns P and Q are located at $(30^\circ$ N, 30° E) and $(30^\circ$ N, 150° W) respectively.

(a) How far apart are the two points, measured along the parallel of latitude 30° N?

(b) Calculate the great circle distance between P and Q.

15. Find the length of the chord that joins

(a) A $(70^\circ$ N, 28° W) and B $(40^\circ$ S, 32° E)

(b) P $(65^\circ$ S, 36° E) and Q $(65^\circ$ S, 76° E)

Chapter Test 10

Take this test as you would take a test in a class. After you are done, check your work against the answers in the back of the book.

In each Question take $\pi = 3.142$ and $R = 6400\ km$. Round off answers to the nearest tenth when necessary.

1. Two towns P and Q are located at (28° N, 23° E) and (28° N, 37° W) respectively. Calculate

(a) the radius of the circle of latitude

(b) the shortest distance between P and Q, measured along the parallel of latitude

2. An aeroplane takes off from a town P located at (40° N, 52° E) and after flying 1,450 kilometres due east, it reaches a town Q. It then flies due north to the North Pole, calculate

(a) the longitude of Q

(b) the distance of Q from the North Pole

3. Three towns P, Q and R are located at (35° N, 75° E), (35° N, 15° E) and (0°, 15° E) respectively. If an aircraft took 5 hours to fly from P to Q along the parallel of latitude, and the 7 hours to fly from Q to R along the meridian, calculate

(a) the total length of the journey

(b) the average speed of the aircraft

Answers to exercises

Chapter 1

Try this

1(a) $\tan x = \frac{3}{4}$, $\tan y = \frac{4}{3}$ (b) $\tan x = \frac{5}{12}$, $\tan y = \frac{12}{5}$

(c) $\tan x = \frac{15}{8}$, $\tan y = \frac{8}{15}$

2. 1.6   3. 38⁰   4. 16.7 m   5. 13.5 cm   6. 36.9⁰

7. 166.6 m   8. 257 m

9(a) $\cos x = \frac{4}{5}$, $\cos y = \frac{3}{5}$ (b) $\cos x = \frac{8}{17}$, $\cos y = \frac{15}{17}$

(c) $\cos x = \frac{5}{13}$, $\cos y = \frac{12}{13}$

10. 0.1857   11. 63⁰   12. 17.3 m   13. 57.8⁰

14(a) $\sin x = \frac{3}{5}$, $\sin y = \frac{4}{5}$ (b) $\sin x = \frac{15}{17}$, $\sin y = \frac{8}{17}$

(c) $\sin x = \frac{12}{13}$, $\sin y = \frac{5}{13}$

15. 0.8181   16. 73⁰   17. 9.6 m   18. 53.1⁰   19. $10\sqrt{3}$

20. $\sin A = \frac{15}{17}$, $\cos A = \frac{8}{17}$ 21. 52.3 m   22. 35.8⁰   23. 6 cm

24. 96.4⁰   25. 6.4 cm   26. 6.2 cm   27. 7.6

28. $a = 4.6$ $B = 70.9⁰$ $C = 49.1⁰$

Exercise 1.1(a)

1(a) 1.192   (b) 5.145   (c) 0.7002   (d) 2.475   (e) 3.376

(f) 0.8632   (g) 3.806   (h) 11.48

2(a) 36.0⁰   (b) 67.0⁰   (c) 85.0⁰   (d) 47.5⁰   (e) 80.6⁰

(f) 52.8⁰   (g) 81.3⁰   (h) 84.2⁰

Exercise 1.1(b)

1(a) 5.1   (b) 12.9   (c)2.9   2(a) 70   (b)1.6   (c) 2.0

Exercise 1.1(c)

1(a) 33.7⁰   (b) 29.7⁰   (c) 71.0⁰

2(a) 24.8⁰   (b) 28.1⁰   (c) 16.3⁰

Exercise 1.1(d)

1. 41.7⁰   2. 6.5 cm   3. 9.7 m   4. 19.6⁰   5. 19.5 cm

6. 3.2 cm

Exercise 1.1(e)

1. 52 m   2. 173.2 m   3. 196.3 m   4. 153.6 m   5. 48.7 m

6. 57.7 m   7. 139.9 m   8. 282.3 m   9. 261.6 m   10. 5645.6 m

11. 44.0 m   12. 86 m   13. 72.5 m   14. 7.7 m   15. 5.4 km

16. 57 m   17. 81 m

Exercise 1.2(a)

1(a) 0.9925   (b) 0.9339   (c) 0.6561   (d) 0.2419   (e) 0.0523

2(a) 0.9980   (b) 0.7934   (c) 0.6115   (d) 0.3714   (e) 0.1840

3(a) 0.9600   (b) 0.8786   (c) 0.7459   (d) 0.4667   (e) 0.0809

4(a) 5.0⁰   (b) 36.0⁰   (c) 56.0⁰   (d) 72.0⁰   (e) 83.0⁰

5(a) 16.4⁰   (b) 26.4⁰   (c) 44.6⁰   (d) 59.2⁰   (e) 79.7⁰

6(a) 17.5^0 (b) 38.6^0 (c) 60.6^0 (d) 78.5^0 (e) 89.4^0

Exercise 1.2(b)

1(a) 5   (b) 1.2   (c) 6.3   (d) 21.5   (e) 48.2^0 (f) 64.8^0

2(a) 50 m   (b) 48.2^0 (c) 24.5

Exercise 1.3(a)

1(a) 0.0349   (b) 0.3907   (c) 0.6561   (d) 0.8387   (e) 0.9877

2(a) 0.0279   (b) 0.2198   (c) 0.6129   (d) 0.8949   (e) 0.9910

3(a) 0.4188   (b) 0.6854   (c) 0.8584   (d) 0.9477   (e) 0.9987

4(a) 5.0^0 (b) 38.0^0 (c) 48.0^0 (d) 70.0^0 (e) 83.0^0

5(a) 0.4^0 (b) 26.4^0 (c) 42.6^0 (d) 74.8^0 (e) 87.2^0

6(a) 6.1^0 (b) 32.1^0 (c) 55.4^0 (d) 67.4^0 (e) 78.5^0

Exercise 1.3(b)

1(a) 4.7   (b) 8.2   (c) 6.6   (d) 11.8   (e) 38.7^0 (f) 40.5^0

2(a) 28.8^0 (b) 53.1^0 (c) 153.2   (d) 26

Exercise 1.3(c)

1. 8.5 m   2. $25\sqrt{3}$ m   3. $24\sqrt{3}$ m   4. 5 m

Exercise 1.3(d)

1(a) 15.7   (b) 51.1^0 (c) 66.4^0 (d) 56.3^0 (e) 10.5

(f) 5.4   (g) 11.8   (h) 11.3   (i) 53.1^0

2(a) 8.1   (b) 9.2   (c) 58^0 (d) 30^0

3. $\tan A = \dfrac{7}{24}$ $\cos A = \dfrac{24}{25}$ 4. $\dfrac{65}{48}$ 5. $\dfrac{161}{289}$ 6. $\dfrac{169}{25}$

7. $a = 12$ $b = 16$ 8. 36 cm

Exercise 1.4

1. 28.7^0 2. 2.6 m   3. 3.2 m   4. 1.7 m   5. 34.5 m   6. 78.8 m

7. 321.5 m   8. 25 km   9. 45 m   10. 30^0

Exercise 1.5

1. 19.3^0 2(i) 67.4^0 (ii) 13 m   3. 48.6^0 4. 32^0

5. 219 m   6. 7 m   7. 31^0 8. 5 m

Exercise 1.6(a)

1. 6 m   2. 13 cm   3. 24 cm   4. 8 cm   5. 6.8 cm   6. 7.4^0

Exercise 1.6(b)

1. 7.9 cm   2. 82.8^0 3. 83.6^0 4. 12 cm   5. 7.1 cm   6. 42.6 cm

Exercise 1.7

1(a) $b = 9.5, c = 12.0, C = 48^0$ (b) $a = 10.3, c = 23.0, B = 30^0$

(c) $a = 13.4, b = 9.8, C = 60^0$ (d) $c = 2.4, A = 25.6^0, B = 133.9^0$

(e) $a = 3.7, B = 95.1^0, C = 36.7^0$ (f) $b = 14.1, c = 8.8, B = 75^0$

(g) $A = 34.8^0, B = 58.8^0, C = 86.4^0$ (h) $A = 81.8^0, B = 60.0^0, C = 38.2^0$

2. 21.3 m   3. 502.7 m   4. 21.8 km   5. 143.3 m   6. 76^0

7. 18.2 cm   8. 211.8 m   9. 165.4 km   10. 583 km

11. 561.1 km   12. 198.5 km

Review exercise 1

1(a) 0.7536   (b) 1.376   (c) 2.145   (d) 3.191

2(a) 0.9890   (b) 0.6820   (c) 0.4695   (d) 0.2045

3(a) 0.4226   (b) 0.8973   (c) 0.9563   (d) 0.9968

4(a) 19.6^0 (b) 36^0 (c) 51.8^0 (d) 71.4^0

5(a) 28.8^0 (b) 44.6^0 (c) 58.3^0 (d) 73.8^0

6(a) 27.2^0 (b) 51.3^0 (c) 61.2^0 (d) 40.9^0

7(a) 3.5   (b) 53.1^0 (c) 11.3   (d) 8.6   (e) 28.8

(f) 26.4^0 (g) 76.2 cm   (h) 68^0 (i) 12.7 m

8. $\sin A = \frac{3}{5}$ $\cos A = \frac{4}{5}$ 9. 2.9 m   10. 34 m   11. 60.9 m

12. 219.8 m   13. 117.7 m   14. 105.9 m   15. 15.7 m   16. 34.8^0

17. 1.5 m   18. 7.9 m   19. 6.4 cm   20(a) 7.8 cm   (b) 100^0

21. 8.6 cm   22. 20.0 cm   23. 37^0

24(a) $b = 2.7, c = 3.1, A = 75^0$ (b) $a = 10.2, B = 66.3^0, C = 52.7^0$

(c) $b = 28.5, B = 108.2^0, C = 41.8^0$ (d) $A = 22.9^0, B = 25.1^0, C = 132^0$

25. 1199.0 m   26. 74.8 m   27. 2.7 minute

Chapter Test 1

1(a) 7.5   (b) 70.5^0 (c) 7.5   2. 52.5 m   3. $\frac{7}{25}$

4(a) 13.1 cm   (b) 6.7 cm   5. 21.7 m   6. 1728.3 km h^{-1}

7. $5\sqrt{6}$ 8. 51.6 m   9(a) 59.1 m   (b) 237.5 m

Chapter 2

Try this

1(a) 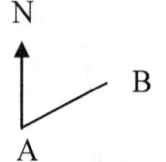   (b)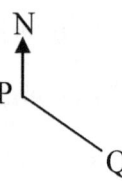

(c) 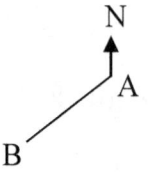   (d)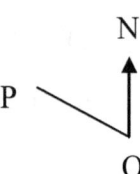

2. 315⁰   3(a) 238⁰   (b) 303⁰   (c) 047⁰   (d) 165⁰

4. 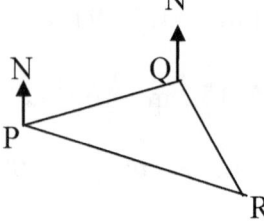   5.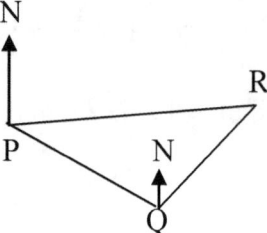

6. 093⁰   7. 17 km, 125⁰

Exercise 2.1

1(a) 253⁰   (b) 234⁰   (c) 288⁰   (d) 317⁰   (e) 030⁰

 (f) 085⁰   (g) 135⁰   (h) 150⁰

2(a) 215⁰   (b) 298⁰   (c) 065⁰   (d) 147⁰

3(a) 206⁰   (b) 263⁰   (c) 307⁰   (d) 283⁰   (e) 045⁰   (f) 103⁰

(g) 127⁰   (h) 135⁰   (i) 180⁰   (j) 270⁰   (k) 000⁰   (l) 090⁰

Exercise 2.2

6. 255^0 7. 085^0 8. 160^0 9. 264^0 10. 030^0 11. 7.5 km

12. 143^0 13. 50 km, 113^0 14. 13 km, 277^0 15. 17 km

16(a) 12.2 km (b) 35^0 (c) 205^0 17. 170 km, 058^0 18. 49 km

19. 8 km   20(a) 14.4 km, 286^0 (b) 4.36 p.m.   21. 106.3 km

22(a) 28.2 km, 318^0 (b) 5.6 km h^{-1}

Review exercise 2

3(a) 266^0 (b) 038^0 (c) 333^0 (d) 134^0

4(a) 223^0 (b) 140^0 (c) 355^0 (d) 079^0

5. 270^0 6. 075^0 7. 096^0 8. 13 km 097^0

9. 17 km 282^0 10. 50 km 253^0

Chapter Test 2

1(a) 288^0 (b) 067^0 2(a) 225^0 (b) 128^0 3. 300^0 4. 260^0

5. 085^0 6. 156 km, 120^0 7. 21 km, 273^0 8. 17 km, 253^0

Chapter 3

Try this

1(a) $\begin{pmatrix} -4 \\ -1 \end{pmatrix}$ (b) $\begin{pmatrix} 0 \\ 3 \end{pmatrix}$ (c) $\begin{pmatrix} 6 \\ 2 \end{pmatrix}$ (d) $\begin{pmatrix} 2 \\ -2 \end{pmatrix}$ (e) $\begin{pmatrix} -4 \\ 0 \end{pmatrix}$

2. $-a = \begin{pmatrix} 7 \\ 3 \end{pmatrix}$ $-x = \begin{pmatrix} 0 \\ -4 \end{pmatrix}$ 3. $\begin{pmatrix} 3 \\ -2 \end{pmatrix}$ 4. $\begin{pmatrix} 2 \\ -3 \end{pmatrix}$ 5. $\begin{pmatrix} -2 \\ 3 \end{pmatrix}$

6. $\begin{pmatrix} 2 \\ 11 \end{pmatrix}$ 7. 10   8. 67^0 9. $\begin{pmatrix} 25 \\ 43.3 \end{pmatrix}$ 10. (5 units, 323^0)

11. $\begin{pmatrix} 6 \\ -4 \end{pmatrix}$ 12. $\begin{pmatrix} 8 \\ -6 \end{pmatrix}$ 13. $x = -5, y = 1$

14(a) not parallel   (b) parallel

15(a) perpendicular   (b) not perpendicular

16. $\overrightarrow{AP}$ is perpendicular to $\overrightarrow{BC}$

Exercise 3.1(a)

1(a) $\overrightarrow{AB} = \begin{pmatrix} 2 \\ 3 \end{pmatrix}$ (b) $\overrightarrow{BC} = \begin{pmatrix} 4 \\ -1 \end{pmatrix}$ (c) $\overrightarrow{CD} = \begin{pmatrix} -2 \\ 4 \end{pmatrix}$ (d) $\overrightarrow{DA} = \begin{pmatrix} -4 \\ 2 \end{pmatrix}$

Exercise 3.1(b)

1. $\begin{pmatrix} -3 \\ -4 \end{pmatrix}$ 2. $\begin{pmatrix} 0 \\ -2 \end{pmatrix}$ 3. $\begin{pmatrix} 5 \\ 4 \end{pmatrix}$ 4. $\begin{pmatrix} 1 \\ -3 \end{pmatrix}$ 5. $\begin{pmatrix} -6 \\ 2 \end{pmatrix}$ 6. $\begin{pmatrix} 7 \\ 0 \end{pmatrix}$

Exercise 3.2(a)

1(a) $\begin{pmatrix} 4 \\ 5 \end{pmatrix}$ (b) $\begin{pmatrix} 2 \\ -4 \end{pmatrix}$ (c) $\begin{pmatrix} 5 \\ -2 \end{pmatrix}$ (d) $\begin{pmatrix} 2 \\ -2 \end{pmatrix}$ (e) $\begin{pmatrix} -2 \\ 2 \end{pmatrix}$ (f) $\begin{pmatrix} -5 \\ 2 \end{pmatrix}$

2(a) $\begin{pmatrix} 3 \\ 4 \end{pmatrix}$ (b) $\begin{pmatrix} 3 \\ 4 \end{pmatrix}$ commutative property

3(a) $\begin{pmatrix} 3 \\ 1 \end{pmatrix}$ (b) $\begin{pmatrix} 3 \\ 1 \end{pmatrix}$ associative property

Exercise 3.2(b)

1(a) $\begin{pmatrix} 1 \\ -1 \end{pmatrix}$ (b) $\begin{pmatrix} -3 \\ -3 \end{pmatrix}$ (c) $\begin{pmatrix} -3 \\ 2 \end{pmatrix}$ (d) $\begin{pmatrix} 2 \\ -3 \end{pmatrix}$ (e) $\begin{pmatrix} 0 \\ 2 \end{pmatrix}$ (f) $\begin{pmatrix} -3 \\ 6 \end{pmatrix}$

2(a) $\begin{pmatrix} -5 \\ 6 \end{pmatrix}$ (b) $\begin{pmatrix} 5 \\ -6 \end{pmatrix}$ Vector subtraction is not commutative

Exercise 3.2(c)

1(a) $\begin{pmatrix} 6 \\ -4 \end{pmatrix}$ (b) $\begin{pmatrix} 5 \\ 0 \end{pmatrix}$ (c) $\begin{pmatrix} 4 \\ 8 \end{pmatrix}$ (d) $\begin{pmatrix} 9 \\ 3 \end{pmatrix}$

(e) $\begin{pmatrix} 4 \\ 6 \end{pmatrix}$ (f) $\begin{pmatrix} 2 \\ -4 \end{pmatrix}$ (g) $\begin{pmatrix} 6 \\ -9 \end{pmatrix}$ (h) $\begin{pmatrix} 6 \\ -9 \end{pmatrix}$

2(a) $\begin{pmatrix} 4 \\ 6 \end{pmatrix}$ (b) $\begin{pmatrix} -5 \\ -7 \end{pmatrix}$ (c) $\begin{pmatrix} -13 \\ 5 \end{pmatrix}$ (d) $\begin{pmatrix} -7 \\ -12 \end{pmatrix}$

3. $\begin{pmatrix} 3 \\ -3 \end{pmatrix}$ 4. $a = 3 \ \ b = 2$ 5. $a = -2 \ \ b = 3$

Exercise 3.3

1. 5   2. 17   3. 15   4. 9.43   5. 7.21   6. 13

Exercise 3.4

1. 53.1^0 2. 68.2^0 3. 51.3^0 4. 51.3^0 5. 36.9^0

Exercise 3.5

1(a) $\begin{pmatrix} 36 \\ 35 \end{pmatrix}$ (b) $\begin{pmatrix} 71 \\ 71 \end{pmatrix}$ (c) $\begin{pmatrix} -4 \\ -3 \end{pmatrix}$

(d) $\begin{pmatrix} -60 \\ 104 \end{pmatrix}$ (e) $\begin{pmatrix} -6 \\ 10 \end{pmatrix}$ (f) $\begin{pmatrix} -11 \\ 11 \end{pmatrix}$

2(a) (13 units, 023^0)   (b) (10 units, 323^0)   (c) (10.3 units, 151^0)

(d) (19.4 units, 214^0)   (e) (17 units, 152^0)

Exercise 3.6

1. $\begin{pmatrix} 0 \\ 6 \end{pmatrix}$ 2. $\begin{pmatrix} 6 \\ 1 \end{pmatrix}$ 3. $\begin{pmatrix} 12 \\ -9 \end{pmatrix}$ 4. $\overrightarrow{AC} = \begin{pmatrix} 5 \\ 1 \end{pmatrix}$ $\overrightarrow{AD} = \begin{pmatrix} 3 \\ -1 \end{pmatrix}$

5. $\overrightarrow{QT} = \begin{pmatrix} -6 \\ 3 \end{pmatrix}$ $\overrightarrow{RS} = \begin{pmatrix} -1 \\ 5 \end{pmatrix}$

Exercise 3.7

1. $\overrightarrow{AB} = \begin{pmatrix} 2 \\ 1 \end{pmatrix}$ $\overrightarrow{BA} = \begin{pmatrix} -2 \\ -1 \end{pmatrix}$ 2. $\begin{pmatrix} 5 \\ -3 \end{pmatrix}$

3. $\overrightarrow{PQ} = \begin{pmatrix} 1 \\ 2 \end{pmatrix}$ $\overrightarrow{PR} = \begin{pmatrix} 8 \\ -2 \end{pmatrix}$ $\overrightarrow{QR} = \begin{pmatrix} 7 \\ -4 \end{pmatrix}$

4. $(3, -2)$ 5. $(0, 7)$ 6. $(4, -3)$

Exercise 3.8

1. $\triangle ABC$ is a right-angled triangle   2. $\overrightarrow{PQ}$ // $\overrightarrow{BC}$ and $\overrightarrow{PQ} = \frac{1}{2}\overrightarrow{BC}$

3. $\triangle ABC$ is an isosceles triangle   4. $x = -2$ $y = 6$

5. $\overrightarrow{PR}$ is perpendicular to $\overrightarrow{SQ}$

Review exercise 3

1(a) $\begin{pmatrix} 1 \\ -3 \end{pmatrix}$ (b) $\begin{pmatrix} -3 \\ -2 \end{pmatrix}$ (c) $\begin{pmatrix} -2 \\ 1 \end{pmatrix}$ (d) $\begin{pmatrix} 1 \\ -2 \end{pmatrix}$ (e) $\begin{pmatrix} -3 \\ -4 \end{pmatrix}$ (f) $\begin{pmatrix} -2 \\ 0 \end{pmatrix}$

2(a) $\begin{pmatrix} -1 \\ -1 \end{pmatrix}$ (b) $\begin{pmatrix} 1 \\ 4 \end{pmatrix}$ (c) $\begin{pmatrix} 4 \\ -3 \end{pmatrix}$ (d) $\begin{pmatrix} -2 \\ 1 \end{pmatrix}$ (e) $\begin{pmatrix} 5 \\ 0 \end{pmatrix}$ (f) $\begin{pmatrix} 1 \\ 2 \end{pmatrix}$

3(a) $\begin{pmatrix} 2 \\ -6 \end{pmatrix}$ (b) $\begin{pmatrix} 6 \\ -12 \end{pmatrix}$ (c) $\begin{pmatrix} -8 \\ 4 \end{pmatrix}$ (d) $\begin{pmatrix} -3 \\ 2 \end{pmatrix}$ (e) $\begin{pmatrix} 6 \\ -3 \end{pmatrix}$ (f) $\begin{pmatrix} -6 \\ 9 \end{pmatrix}$

4(a) $\begin{pmatrix} 0 \\ 11 \end{pmatrix}$ (b) $\begin{pmatrix} -11 \\ 0 \end{pmatrix}$ (c) $\begin{pmatrix} -6 \\ -7 \end{pmatrix}$ (d) $\begin{pmatrix} -12 \\ 7 \end{pmatrix}$

(e) $\begin{pmatrix} -17 \\ -7 \end{pmatrix}$ (f) $\begin{pmatrix} -9 \\ -4 \end{pmatrix}$ (g) $\begin{pmatrix} 16 \\ 11 \end{pmatrix}$

5(a) 5   (b) 17   (c) 10   (d) 13

6(a) 37^0 (b) 22^0 (c) 61^0 (d) 53^0

7(a) $\begin{pmatrix} 15 \\ 13 \end{pmatrix}$ (b) $\begin{pmatrix} 11 \\ -8 \end{pmatrix}$ (c) $\begin{pmatrix} -6 \\ -8 \end{pmatrix}$

8(a) (13 units, 337^0)   (b) (10 units, 143^0)

(c) (17 units, 208^0)   (d) (5 units, 053^0)

9. $\begin{pmatrix} 5 \\ -3 \end{pmatrix}$ 10. $\begin{pmatrix} 8 \\ -6 \end{pmatrix}$ 11. $\overrightarrow{AB} = \begin{pmatrix} 7 \\ -5 \end{pmatrix}$ $\overrightarrow{BA} = \begin{pmatrix} -7 \\ 5 \end{pmatrix}$

12. (4, - 3)   13. (- 3, 1)   14(a) not parallel   (b) parallel

15(a) perpendicular   (b) not perpendicular

16(a) $\overrightarrow{AB}$ // $\overrightarrow{DC}$ and $\overrightarrow{AB} = \overrightarrow{DC}$

   (b) $\overrightarrow{AC}$ is perpendicular to $\overrightarrow{BD}$ $\overrightarrow{AC} = \overrightarrow{BD}$

Chapter Test 3

1(a) $\begin{pmatrix} 1 \\ 0 \end{pmatrix}$ (b) $\begin{pmatrix} -21 \\ 13 \end{pmatrix}$

2(a) $\overrightarrow{PQ} = \begin{pmatrix} 6 \\ -8 \end{pmatrix}$ $\overrightarrow{QR} = \begin{pmatrix} -6 \\ -2 \end{pmatrix}$ $\overrightarrow{PR} = \begin{pmatrix} 0 \\ -10 \end{pmatrix}$

   (b) ΔPQR is isosceles

3. $k = 2, m = 1$

4(a) $\overrightarrow{AB} = \begin{pmatrix} 8 \\ -15 \end{pmatrix}$, $\overrightarrow{BC} = \begin{pmatrix} -5 \\ 12 \end{pmatrix}$ (b) $\begin{pmatrix} 3 \\ -3 \end{pmatrix}$ (c) (4.2 km, 135°)

5(a) $\overrightarrow{AB} = \begin{pmatrix} 3 - x \\ 2 - y \end{pmatrix}$, $\overrightarrow{DC} = \begin{pmatrix} 8 \\ 4 \end{pmatrix}$, $x = -5, y = -2$ (b) 14.8

Chapter 4

Exercise 4.2

1(d) $\angle P_1 P_2 P_3 = 58°$ 2(d) $|P_1 P_2| = 3.9$ cm   3(b) $\angle BPD = 30°$

4(d) 2.7 cm   5(b) 6.8 cm   (c) 22 cm^2

Exercise 4.3

3(c) 29 m   5(c) 15 km

Review exercise 4

1. 4.3 cm   2. 2 cm   3. 15°   4. 2.3 cm

5. 90°,   Right-angled isosceles triangle   2.9 cm

6. 60° 7. 7.2 cm   8. 90° 3 cm   9. Rhombus with sides 3.5 cm

Chapter Test 4

1(b) 90° (f) 1.6 cm

Chapter 5

Try this

1. $\frac{1}{10}$ 2. $\frac{1}{2}$ 3(a) (i) 0   (ii) 1   (b) (i) $\frac{2}{5}$ (ii) $\frac{7}{20}$ (iii) $\frac{1}{4}$ (iv) 1

4(a) $\frac{1}{3}$ (b) $\frac{2}{3}$ 5(a) 0.26   (b) 0.57   (c) 0.71   6. $\frac{2}{7}$ 7. $\frac{1}{36}$

8. $\frac{4}{13}$ 9. $\frac{5}{18}$ 10(a) $\frac{1}{6}$ (b) $\frac{1}{2}$ (c) $\frac{5}{6}$ 11. $\frac{13}{28}$

Exercise 5.1

1. $\frac{1}{2}$ 2. $\frac{1}{3}$ 3. $\frac{2}{5}$ 4. $\frac{1}{2}$ 5. $\frac{3}{10}$ 6. $\frac{1}{2}$ 7. $\frac{5}{12}$

8. 15   9. $\frac{3}{13}$ 10. $\frac{1}{2}$ 11. 30   12. 600

Exercise 5.2

1(a) $\frac{1}{5}$ (b) $\frac{1}{2}$ (c) $\frac{3}{10}$ 2. $\frac{2}{15}$ 3(a) $\frac{1}{6}$ (b) $\frac{17}{24}$ (c) $\frac{1}{2}$

4(a) $\frac{1}{4}$ (b) $\frac{13}{40}$ 5(a) $\frac{3}{8}$ (b) $\frac{3}{10}$ 6. $\frac{13}{20}$

Exercise 5.3(a)

1. $\frac{1}{3}$ 2. $\frac{2}{3}$ 3. $\frac{2}{3}$ 4. $\frac{3}{5}$ 5. $\frac{1}{10}$ 6. $\frac{2}{3}$

7. $\frac{3}{4}$ 8. $\frac{1}{2}$ 9. $\frac{2}{13}$ 10. $\frac{2}{13}$ 11. $\frac{11}{20}$ 12. $\frac{13}{20}$

Exercise 5.3(b)

1. $\frac{1}{12}$ 2. $\frac{1}{16}$ 3. $\frac{9}{25}$ 4. $\frac{6}{25}$ 5. $\frac{2}{5}$

6. $\frac{1}{2}$ 7. $\frac{6}{25}$ 8. $\frac{1}{64}$ 9. $\frac{23}{45}$ 10. $\frac{29}{60}$

Exercise 5.3(c)

1. $\frac{3}{4}$ 2. $\frac{11}{36}$ 3. $\frac{7}{12}$ 4. $\frac{8}{13}$ 5. $\frac{7}{13}$ 6. $\frac{4}{13}$

7. $\frac{31}{40}$ 8. $\frac{13}{20}$ 9. $\frac{7}{10}$ 10. $\frac{43}{60}$ 11. $\frac{17}{20}$ 12. $\frac{5}{6}$

Exercise 5.3(d)

1. $\frac{1}{17}$ 2. $\frac{13}{51}$ 3. $\frac{33}{95}$ 4. $\frac{4}{15}$ 5. $\frac{4}{51}$

6. $\frac{2}{21}$ 7. $\frac{1}{21}$ 8. $\frac{28}{153}$ 9. $\frac{84}{325}$ 10. $\frac{10}{21}$

Exercise 5.4

1(a) $\frac{1}{6}$ (b) $\frac{1}{4}$ 2(a) $\frac{1}{4}$ (b) $\frac{3}{4}$ 3(a) $\frac{11}{36}$ (b) $\frac{1}{9}$ (c) $\frac{5}{12}$

4(a) $\frac{1}{6}$ (b) $\frac{1}{2}$ 5. $\frac{12}{35}$ 6. $\frac{8}{15}$ 7. $\frac{18}{35}$ 8. $\frac{10}{33}$ 9. $\frac{7}{15}$ 10. 0.855

Review exercise 5

1. $\frac{4}{11}$ 2. $\frac{18}{23}$ 3. $\frac{1}{2}$ 4. $\frac{1}{5}$ 5. $\frac{3}{13}$

6. $\frac{2}{5}$ 7. $\frac{3}{10}$ 8. $\frac{1}{13}$ 9. $\frac{1}{5}$ 10. $\frac{2}{13}$

11. $\frac{11}{36}$ 12. $\frac{4}{13}$ 13. $\frac{1}{12}$ 14. $\frac{1}{16}$ 15. $\frac{4}{25}$

16. $\frac{13}{24}$ 17. $\frac{1}{17}$ 18. $\frac{7}{145}$ 19. $\frac{6}{25}$ 20. $\frac{8}{15}$

Chapter Test 5

1. $\frac{23}{50}$ 2. $\frac{2}{3}$ 3. $\frac{7}{20}$ 4. $\frac{7}{10}$ 5. $\frac{11}{26}$ 6. $\frac{5}{18}$

7. $\frac{4}{25}$ 8. $\frac{3}{4}$ 9. $\frac{7}{12}$ 10. $\frac{15}{28}$ 11. $\frac{5}{9}$ 12. $\frac{7}{15}$

Chapter 6

Try this

1.

Marks	Tally	Frequency
25	///	4
26	ᵁᴴᵀ//	7
27	ᵁᴴᵀ////	9
28	ᵁᴴᵀ	5
29	///	3
30	//	2

2.

Ages(in years)	Tally	Frequency
10 – 19	///	3
20 – 29	ᵁᴴᵀ	5
30 – 39	ᵁᴴᵀ ////	9
40 – 49	ᵁᴴᵀ //	7
50 – 59	////	4
60 – 69	//	2

5.

Height less than	Cumulative frequency
114.5	2
119.5	6
124.5	12
129.5	22
134.5	30
139.5	36
144.5	40

7. 6   8. 6.3   9. 7.1   10. 33.5

11(a) 16   (b) 4, 5   12(a) 3   (b) 8.5   (c) 4

13(a) 4 th   (b) $5\frac{1}{2}$ th   14. 16   15. $Q_1 = 18$ $Q_2 = 31$

16. 61.5   17. $Q_2 = 71.0$ $Q_1 = 65.5$ $Q_3 = 75.5$

18. 34   19. 18   20. 7   21. 1.7   22. 2.1   23. 3.5

24. 1.9   25. 6   26(b) positive correlation   (c) GH¢ 63.00

Exercise 6.1

1.

Ages(in years)	Tally	Frequency
7	/	1
8	///	3
9	////-	5
10	////	6
11	/////	7
12	///////	8
13	///	3
14	//	2
15	/	1

2.

Marks	Tally	Frequency
5	////	4
6	ЦЊ	5
7	ЦЊ ////	9
8	ЦЊ //	7
9	///	3
10	//	2

3.

Height	Tally	Frequency
151 – 155	////	4
156 – 160	ЦЊ	5
161 – 165	ЦЊ ////	9
166 – 170	ЦЊ //	7
171 – 175	///	3
176 – 180	//	2

4.

Ages	Tally	Frequency
21 – 25	////	4
26 – 30	////	4
31 – 35	ЦЊ ////	9
36 – 40	ЦЊ ////	9
41 – 45	ЦЊ //	7
46 – 50	////	4
51 – 55	//	2
56 – 60	/	1

5.

Marks	Tally	Frequency
0 – 9	//	2
10 – 19	////	4
20 – 29	////̸	5
30 – 39	////̸ //	7
40 – 49	////̸ ///	8
50 – 59	////̸ ////	9
60 – 69	////̸ /	6
70 – 79	////	4
80 – 89	///	3
90 – 99	//	2

Exercise 6.2(b)

3(a) 1,400   4. 24   7(a) GH¢ 10,125.00   (b) $33\frac{1}{3}$%

Exercise 6.4(a)

1(a) 15.8   (b) 56   (c) 19.1   (d) 27   2. 26.7   3. 15.55

4(a) 6.33   (b) 38.6   5. 9   6. 15   7. 7   8. 6.52

9. 16.5   10. 16.2   11. 19.7   12. 33.1

Exercise 6.4(b)

1(a) 5   (b) 15   (c) 3   (d) 21 26   (e) no mode

2. 4   3. 11   4. 20 – 24   5. 11 – 15

Exercise 6.4(c)

1(a) 6   (b) 19   (c) 4   (d) 6.5

2(a) 16   (b) 29   (c) 13   (d) 12

3. 19   4. 2   5(a) 4 15   (b) 26 36   (c) 9 21

Exercise 6.4(d)

1. 147   2. 19   3. 44.5   4. 53.5 34.5   67 5. 44   33   54.5

6. 56.5 cm   86 pegs   7. 15   8. 49kg   9

Exercise 6.5

1(a) 9   (b) 12   2(a) 8   (b) 8.5   3(a) 3.5   (b) 4.5

4. 4.5   5. 9   6(a) 3.5   (b) 8   (c) 2.28   (d) 1.44

7. 3.488   8. 6.352   9. 5.384   10. 1.16   11(a) 2.71 (b) 5.8

12. 1.42   13(a) 37.6   (b) 155.6   (c) 27.96

Exercise 6.6

1(a) positive correlation   (d) 94

2(a) 82   (b) positive correlation

3(b) negative correlation   (d) 65 hours

Review exercise 6

1.

Ages(in years)	Tally	Frequency
12	////	4
13	////	4
14	////-///	8
15	////-///	8
16	////	4
17	//	2

2.

Length(cm)	Tally	Frequency
11 – 15	///	3
16 – 20	////	4
21 – 25	~~HHT~~ ~~HHT~~	10
26 – 30	~~HHT~~ ////	9
31 – 35	~~HHT~~ //	7
36 – 40	////	4
41 – 45	//	2
46 – 50	/	1

5(a) 8.3   9   9   (b) 17.25   16   17   (c) 25.2   26   26

6. 34   7. 43   8. 6   9. 57   10. 15.875   11. 22   12. 20.75

13. 71.2   68   74.3   14. 47.7   36   60.3   15. 42.5   9.8

16(a) 9   (b) 12   (c) 19   17(a) 10   (b) 12   18(a) 5   (b) 6

19. 3.67   20. 2.4   21. 0.95   22. 4.64

23(a) 2.45   (b) 4.465   (c) 11.74   24(a) 27   (b) 7.6   (c) 91.4

25(b) positive correlation   26(b) positive correlation   (d) 69

Chapter Test 6

1(a)

Marks	Frequency
5	3
6	6
7	6
8	9
9	4
10	2

(b) 8   (c) 7   (d) 7.37

2(a)

Heights	Frequency
140 – 144	1
145 – 149	1
150 – 154	5
155 – 159	3
160 – 164	8
165 – 169	6
170 – 174	4
175 – 179	2

(c) 163

4. 19.4 years   5(b) (i) 47.5   (ii) 26   6. Range = 8   SIQR = 2

7. 5.03   8. 2.41   9. 11.8   10(b) negative correlation   (c) 8

Chapter 7

Try this

1(b) (i) (3, -1)   (iii) $B_1(4, -2)$, $C_1(2, -3)$

 (c) (ii) $A_2(-3,1)$, $B_2(-4,2)$, $C_2(-2,3)$

 (d) (iii) $A_3(1,3)$, $B_3(2,4)$, $C_3(3,2)$

 (e) (iii) $A_4(-1,3)$, $B_4(-2,4)$, $C_4(-3,2)$

 (f) (ii) $A_5(1, -1)$, $B_5(0, -2)$, $C_5(2, -3)$

 (g) (ii) $A_6(-3, -3)$, $B_6(-4, -4)$, $C_6(-2, -5)$

2(a) (2,3)   (b) (-2, -3)   (c) (-3, 2)   (d) (3, -2)

3(a) (-4,1)   (b) (4,1)   (c) (-2,3)   (d) (-2, -3)

4(b) (i) $A'(-2,2)$ (c) (ii) $A''(-2,-2)$, $B''(-2,-3)$, $C''(-3,-2)$

(d) (ii) $A'''(2,-2)$, $B'''(3,-2)$, $C'''(2,-3)$

5(a) (-1,-3) (b) (3, -1) (c) (1,3)

6(b) (i) $A'(3,3)$ (ii) $B'(4,3)$, $C'(4,4)$ (d) (ii) $\begin{pmatrix} -3 \\ 2 \end{pmatrix}$

7(a) (3,4) (b) (2, -2) (c) (5, -1) 8(b) 2 9. 2

11. $A'(6,-3)$, $B'(-3,3)$ $C'(-6,-9)$ 12. 4 cm^2 13. 2.8 cm

Exercise 7.1

1(a) (3, -5) (b) (- 4.7) (c) (5, 2) (d) (0,6) (e) (-3, -2)

2(a) (-5,7) (b) (7, - 4) (c) (-2, - 5) (d) (-6, 0) (e) (2, 3)

3(a) (3, 2) (b) (4, -3) (c) (-2, 5) (d) (0 , 7) (e) (- 6, -1)

4(a) (-7, -5) (b) (-3, 2) (c) (4, -3) (d) (5, 4) (e) (2, 0)

5(a) (0, -3) (b) (6, -1) (c) (-2,5)

6(a) (5, -4) (b) (3, -10) (c) (-8, 0)

7(a) x – axis (b) y = x (c) y – axis (d) y = - x

8(a) x = 3 (b) y = 2 (c) x = 6 (d) y = 1 (e) y = - 2

Exercise 7.2

1(a) (-5,3) (b) (0, -1) (c) (3, -2) (d) (- 4, 0) (e) (3, 5)

2(a) (2, -3) (b) (-4, 2) (c) (3, 1) (d) (- 2, -1) (e) (0, 4)

3(a) (2, -3) (b) (- 3, 0) (c) (-4, 5) (d) (3, 6) (e) (0, - 3)

Exercise 7.3

1(a) (i) (4,5)   (ii) (-1,2)   (iii) (3,0)   (iv) (1,2)

  (b) (i) (1,7)   (ii) (-4,4)   (iii) (0,2)   (iv) (-2, 4)

  (c) (i) (5, 2)   (ii) (0, -1)   (iii) (4, -3)   (iv) (2, -1)

  (d) (i) (-1, -1) (ii) (-6, -4)   (iii) (-2, -6) (iv) (- 4, -4)

2(a) (i) (3,2)   (ii) (7,0)   (iii) (8, 5)   (iv) (7, 3)

  (b) (i) (-6, 1)   (ii) (-2, -1)   (iii) (-1, 4)   (iv) (-2, 2)

  (c) (i) (-8, -6) (ii) (-4, -8)   (iii) (-3, -3) (iv) (-4, -5)

  (d) (i) (2, -4)   (ii) (-2, -6)   (iii) (-3,=1)   (iv) (-2, -3)

3(a) $\begin{pmatrix} -2 \\ 2 \end{pmatrix}$ (b) $\begin{pmatrix} 2 \\ 4 \end{pmatrix}$ (c) $\begin{pmatrix} -5 \\ -2 \end{pmatrix}$ (d) $\begin{pmatrix} 9 \\ -4 \end{pmatrix}$ (e) $\begin{pmatrix} -4 \\ -3 \end{pmatrix}$

Exercise 7.4

3(a) 2   (b) 5   (c) 8   4(a) $\frac{1}{3}$ (b) 4   (c) 15   5(a) – 2   (b) 10

6(a) (4, 6)   (b) (-2, 3)   (c) (-3, 4)   (d) (0, 2)   (e) (-3, -6)

7(a) (1, 2)   (b) (-17, 0)   (c) (5, 5)   (d) (-11, -8)

8(a) (1, -2)   (b) (-8, 16)   (c) (10, 1)   (d) (13, 0)

9(a) – 2, (b) origin   10(a) 3 (b) (2, 2)

Exercise 7.5(a)

1. 1 : 3   2. $7\sqrt{3}$ 3. 5 cm   4. 4 cm   5. 108 cm^2 6. 22.5 cm^2

7. 3 cm^2 8. 8 m^2 9. GH¢ 400   10. 900

Exercise 7.5(b)

1. $3 : 1$ 2. 162 cm^3 3. 6 cm   4. 5346 cm^3 5. 12 cm

6. 14.3 cm   7. 4 m   8. 54 cm^3 9. $0.25 \text{ cm}^3 \text{ s}^{-1}$ 10. 8

Review exercise 7

1 (i) (a) (-2,4)   (b) (2, -4)   (c) (4,2)   (d) (-4, -2)

(ii) (a) (3, -5)   (b) (-3, 5)   (c) (- 5, -3)   (d) (5, 3)

(iii) (a) (-3, -2)   (b) (3, 2)   (c) (-2, 3)   (d) (2, -3)

2(ii) (a) (9, 2)   (b) (-1, 2)   (c) (-3, -8)   (d) (-3, 2)

(ii) (a) (2, -5)   (b) (-8, -5)   (c) (4, -1)   (d) (4, 9)

(iii) (a) (7, -2)   (b) (-3, -2)   (c) (-1, -4)   (d) (-1, 6)

3(a) y – axis   (b) x – axis   (c) y = x   (d) y = -x

4(a) x = 2   (b) y = 3   (c) x = -2   (d) y = -1

5(i) (a) (0, -3)   (b) (3, 0)   (c) (0, 3)

(ii) (a) (3, 4)   (b) (-4, 3)   (c) (-3, -4)

(iii) (a) (-4, -5)   (b) (5, -4)   (c) (4, 5)

6(i) (a) (1, 0)   (b) (0, -1)   (c) (-1, 0)

(ii) (a) (3, 4)   (b) (4, -3)   (c) (-3, -4)

(iii) (a) (-3, 5)   (b) (-5, -3)   (c) (3, -5)

7(i) (a) (5, 1)   (b) (-2, 2)   (c) (0, -4)   (d) (7, -3)

(ii) (a) (1, 6)   (b) (-6, 7)   (c) (-4, 1)   (d) (3, 2)

(iii) (a) (0, -2)   (b) (-7, -1)   (c) (-5, -7)   (d) (2, -6)

8(a) $\begin{pmatrix} 1 \\ -3 \end{pmatrix}$ (b) $\begin{pmatrix} -1 \\ 2 \end{pmatrix}$ (c) $\begin{pmatrix} -3 \\ -1 \end{pmatrix}$ (d) $\begin{pmatrix} -3 \\ 3 \end{pmatrix}$

9(i) (a) (4, 8)   (b) (1, 2)   (c) (-3, -6)

(ii) (a) (-8, 12)   (b) (-2, 3)   (c) (6, -9)

(iii) (a) (-12, -4)   (b) (-3, -1)   (c) (9, 3)

10(f) Rotation about the origin through 180^0

11(a) ABC is a right – angled triangle   (d) Reflection in the x – axis

12(d) $T = \begin{pmatrix} -3 \\ -3 \end{pmatrix}$ (g) Reflection in the line y = -3

13 594 cm^2 14. 1.4 cm   15. 211.2 cm^2

16. 2 cm   17. 0.8 m^3 18. 62. 5 cm

Chapter Test 7

1(a) $A'(2,-1)$, $B'(5,-4)$, $C'(1,-4)$

   $A''(-2,-1)$, $B''(-5,-4)$, $C''(-1,-4)$

(c) Rotation about the origin through 180^0

2(b) (i) (1,1)   (ii) 2

3(a) $A'(-6,8)$, $B'(-8,4)$, $C'(-4,4)$

(d) $A''(6,-8)$, $B''(8,-4)$, $C''(4,-4)$

(e) Reflection in the x- axis

4(a) (ii) $A_1(1,-1)$, $B_1(4,-1)$, $C_1(4,-3)$

   (iii) $A_2(1,7)$, $B_2(4,-7)$, $C_2(4,-5)$

   (iv) Reflection in the y- axis

(b) (i) $A_3(-1,-1)$, $B_3(-4,-1)$, $C_3(-4,-3)$

(iii) Reflection in the y- axis

(c) $A_4(-6,1)$, $B_4(0,1)$, $C_4(0,5)$

5. 20.0 cm^2 6. 27 litres   7. 80 m^2

Chapter 8

Try this

1. 6   2. 4.5 cm^2 3. 51.96 cm^2 4. 84 cm^2, 38 cm 5. 108 cm^2

6. 103.92 cm^2 7. 60 cm^2 8(a) 198 cm   (b) 22 cm

9. 5.6 m   10. 58.7 cm   11. 45^0 12(a) 78.6 cm   (b) 201.1 m

13. 2.8 cm   14. 298.6 cm^2 15. 25.14 cm^2 16. 60^0

Chapter 8.1(a)

1. 11 cm   2. 7 cm   3. 16 cm   4. 57 cm   5. 10.2 cm^2

6. 17.5 cm^2 7. 3.36 m^2 8. 54 m^2 9. 12 cm^2 10. 60 cm^2

11. 20 m^2 10 m   12. 8.7 cm   43.3 cm^2 13. 4.8 cm

14. 108 cm^2 14.4 cm

Exercise 8.1(b)

1. 18.7 cm^2 2. 26.8 cm^2 3. 12 cm^2 4. 12.2 cm^2 5. 11.3 m^2

6. 58 m^2 7. 6.4 cm   8. 5 cm   9. 60.1^0

10. 12 cm   11. 8 cm   12. 13 cm

Exercise 8.2(a)

1(a) 22 cm   30 cm^2 (b) 30 m   56 m^2 (c) 48 mm   13.5 mm^2

(d) 20 m   24.36 m^2

2(a) 32 cm   64 cm^2 (b) 60 cm^2 225 cm^2 (c) 14.0 m 12.25 m^2

(d) 28.8 cm   51.84 cm^2 3. 5 cm   4. 9 cm   5, 8 m   6. 13 m   9 m

7. 60 cm^2 8. 180 cm^2 9. 50.9 cm   162 cm^2 10. 15 m

Exercise 8.2(b)

1. 40 cm^2 2. 24 cm^2 3. 7.8 m^2 4. 6 cm   5. 3 cm   6. 9 cm

7. 4 m   8. 4 m   9. 114 cm^2 10. 65 cm^2

Exercise 8.2(c)

1. 26.3 cm^2 2. 65.6 m^2 3. 103.2 cm^2 4. 3 cm   5. 30^0 6. 60^0

Exercise 8.2(d)

1. 22 cm^2 2. 1.2 m   3. 70 cm^2 4. 6 cm   5. 6 cm   6. 42 m

7. 55 cm   8. 4 cm   9. 174 cm^2 54 cm   10. 144 cm^2 54 cm

11. 90 m^2 40 m

Exercise 8.3(a)

1(a) 88 cm   (b) 19.8 m   (c) 47.1 m   (d) 25.8 km

2(a) 13.2 m   (b) 92.4 mm   (c) 37.7 cm   (d) 282.9 km

3(a) 7 cm   (b) 2.45 km   (c) 4.55 m   (d) 11.14 cm

4(a) 1.02 km   (b) 1.27 m   (c) 1.59 cm   (d) 5 m

5. 200   6. 7.13 km   7. 471 m   8. 12

9. 1.4 m   10. 140   11. 16.5 m

Exercise 8.3(b)

1(a) 4.7 m   (b) 94.3 mm   (c) 7.9 m   (d) 31.4 cm

2(a) 74.5^0 (b) 108.1^0 (c) 59.7^0

3. 31.8 m   4. 12 cm

5(a) 28.7 cm   (b) 27.9 cm   (c) 64 m

6. 55.7 cm   7. 51.9 cm

Exercise 8.3(c)

1(a) 19.6 m^2 (b) 98.6 cm^2 (c) 7546 mm^2 (d) 8.0 km^2

2(a) 7.9 cm   (b) 7.7 m   (c) 12.6 mm   (d) 5.0 cm

3. 162.9 m 1314.3 m^2 4. 42 cm^2 5. 20.6 cm^2 6. 21.4 cm^2

7. 13.9 m^2 8. 4.9 m   9. 5 cm   10. GH¢ 3457.14

11(a) 48.9 cm   (b) 873.1 m   (c) 29.5 m

Exercise 8.3(d)

1(a) 22 cm^2 (b) 814 mm^2 (c) 188.6 m^2 (d) 1.54 m^2

2. 176 cm^2 3. 15.9 m^2 4. 10054.4 m^2

5. 111.1 mm^2 6. 47.1 cm^2 7. 770 cm^2

Exercise 8.3(e)

1(a) 1640 mm^2 (b) 462 cm^2 (c) 4.7 m^2 (d) 33.9 cm^2

2. 9.8 cm   3. 15.9 cm   4. 15.1 cm   5. 63^0

6(a) 3.4 m^2 (b) 23.1 cm^2 7. 29.7 cm^2 8. 11.5 cm

Review exercise 8

1(a) 30 m^2 (b) 45 cm^2 (c) 52 m^2

(d) 60 m^2 (e) 100 cm^2 (f) 120 m^2

2. 4.8 cm   3(a) 47.5 m^2 , 31.1 m   (b) 96 m^2, 43m

4. 11 cm   5. 9 m   10. 1073 m^2 7. 29 m   8. 196 cm^2

9. 11.8 m   10. 33.75 cm^2 11. 49 cm^2

12(a) 24 m^2 (b) 84 cm^2 (c) 103.9 m^2

13(a) 33 cm^2 (b) 58.5 m^2 (c) 84 cm^2

14(a) 8.8 m   (b) 20.1 cm   (c) 37.7 mm

15(a) 132 cm   (b) 28.3 m   (c) 45.3 cm

16. 4.4 cm   17. 85.9⁰   18. 64 cm

19(a) 75.5 m^2 (b) 98.6 cm^2 (c) 162.9 cm^2

20. 330 cm^2 21. 18.9 m   28.3 m^2

22. 17.8 cm   25.1 cm^2 23. 397.1 m   7964.3 m^2

24. 152.6 m^2 46.9 m   25. 23.6 cm^2 26. 135⁰

27. 8 cm   28. 10.4 cm   29. 22.1 m^2 30. 154.2 m^2

Chapter Test 8

1(a) 176 cm^2 52 m   (b) 131.6 cm^2 44.9 cm

2(a) 96 cm   (b) 109.0 cm^2 3. GH¢ 135   4. 3 m   5. 8 kg

6. 7 cm   7(a) 7964.3 m^2 (b) 13   8. 7.07 cm^2

9(a) 85⁰   (b) 18.8 cm   10(a) 21 m   (b) 131 m^2

Chapter 9

Try this

1. 300 cm^2 2. 176 cm^2 3. 450 cm^3 4. 258 cm^2 5. 105 cm^3

6. 198 cm^2 7. 842.5 cm^2 8. 843.5 cm^3 9. 204.3 cm^2

10. 282.8 cm^2 11. 1139.5 cm^3 12. 2.4 m 13. 39.7^0

14. 67.4^0 76^0 15. 411.5 cm^2 16. 69.8 cm^3

17. 24.6 m^2 11.5 m^3 18. 46.0 m^3 73.9 m^2

19. 7845 cm^3 20. 659.8 cm^2 21. 1.49 m^3

Exercise 9.1

1. 360 cm^2 2. 180 cm^2

3(a) 210 cm^2 (b) 520 cm^2 (c) 600 cm^2

4. 350 cm^2 5. 440 cm^2 6. 160 cm^2 7. 12 cm 8. 5 cm

9(a) 780 cm^3 (b) 144 cm^3 (c) 222.5 cm^3

10. 18 cm 11. 96 cm^3 12. 126 cm^3

13. 1000 cm^2 14. $16 \text{ cm}^3 \text{ s}^{-1}$ 15. 162 cm^3

16(a) 73.9 cm^2 34.6 cm^3 (b) 516 cm^2 495 cm^3

(c) 340.7 cm^2 249.4 cm^3

Exercise 9.2

1. 180 cm^2 2. 216 cm^3 3. 384 cm^2 4. 729 cm^3 5. 10 cm

6. 800 g 7. 5 hours 8. 2 cm 9. 5 cm 6 cm 7 cm

10. 29 cm 11. 1122.4 cm^3 12. GH¢ 35.25

13. 24 cm   14. 16 cm   15. 12.5 litres

Exercise 9.3

1(a) 132 cm^2 (b) 440 m^2 (c) 66 cm^2 (d) 62.9 m^2

2. 352 cm^2

3(a) 313 .3 cm^2 (b) 880 cm^2 (c) 377.1 cm^2 (d) 2376 m^2

4. 88 cm^2 5. 1571.4 m^2 6. 35.2 cm^2

7. GH¢ 3300   8. GH¢ 923.99

9(a) 301.6 cm^3 (b) 2.0 m^3 (c) 1939.1 mm^3 (d) 243.4 cm^3

10. 4888.5 m^3 11. 1.2 cm   14.3 cm^3 12. 5655.6 cm^3

13. 4.9 litres   14. 8.5 m^3 15. 5.0 cm

Exercise 27.4

1(a) 150.9 cm^2 (b) 204.3 cm^2 (c) 52.5 cm^2

2(a) 66 cm^2 (b) 1131.4 cm^2 (c) 301.7 cm^2

3. 462 cm^2 19.8 cm   4. 7.9 m   5. 28 m

6(a) 75.4 cm^3 (b) 20 cm^3 (c) 50.3 cm^3 (d) 8.5 cm^3

7. 1.6 cm   2.4 cm^3 8. 3.5 cm   9. 5.1 m^3 23.9 m^2

10. 204.2 cm^3 11. 37.7 cm^3 12. 3850 cm^2 1230 cm^3

13. 9.5 cm   14. 7.6 m   15. 9.6 cm

Exercise 9.5(a)

1. 70 cm   2. 13 m   3. 7 cm   4. 17.0 m   5. 56,3^0 6. 63.4^0

7. 70.6^0 8. 45^0 9. 12 cm   10. 12.7 cm

Exercise 9.5(b)

1. 75.1^0 68.2^0 2. 35.8^0 37.1^0 3. 62.3^0 73.7^0 4. 69.3^0

Exercise 9.5(c)

1. 208.2 cm^2 2. 471.3 cm^2 3. 853.2 cm^2 4. 1246.8 m^2

5. 219.6 cm^2 6. 225.2 cm^3 7. 22.0 cm^3 8. 360 cm^3

9. 296.4 m^3 10. 288.7 m^3 11. 6 m 46.7^0 12. 15 m 46.8^0

Exercise 9.6

1(a) 804.6 cm^2 2145.5 cm^3 (b) 201.4 m^2 268.2 m^3

2(a) 452.6 cm^2 905.1 cm^3 (b) 98.6 m^2 92.0 m^3

3(a) 56.6 m^2 84.9 m^2 56.6 m^3

  (b) 509.1 cm^2 763.7 cm^2 1527.4 cm^3

4(a) 308 cm^2 462 cm^2 718.7 cm^3

  (b) 39.3 m^2 58.9 m^2 32.7 m^3

5. 10.1 cm 6. 1.5 cm 7. 57.9 cm^3 8. 314.3 cm^2 9. 452.6 cm^2

10. 21 cm 11. 25.8 cm^2 12. 681 m^2 13. 5.4 cm 14. 2.0 cm

Exercise 9.7

1(a) 320 m^2 678.9 m^3 (b) 339.4 cm^2 640 cm^3

  (c) 533.7 cm^2 1131.4 cm^3

2(a) 432 cm^2 1728 cm^3 (b) 447.2 cm^2 1490.7 cm^3

  (c) 432 m^2 857.2 m^3

3(a) 150.9 m^2 189.6 m^3 (b) 63.2 cm^2 56.6 cm^3

(c) 47.1 m^2 37.7 m^3

4 (a) 48 cm^2 64 cm^3 (b) 17.9 cm^2 11.9 cm^3

(c) 48 m^2 31.7 m^3

5(a) 2413.6 m^2 5732.6 m^3 (b) 8568.2 cm^2 1990.5 cm^3

(c) 5798.6 cm^2 1395.4 cm^3

6(a) 127.9 m^2 1456 m^3 (b) 1428.6 cm^2 2105.2 cm^3

(c) 464 m^2 448 m^3

Exercise 9.8

1. 103.7 cm^2 2. 4189.3 cm^3

3(a) 169.7 m^3 (b) 612.7 cm^2 (c) 1331 cm^3

4(a) 1335.7 cm^3 (b) 1810.3 cm^3 (c) 2877.0 cm^3

5. 1458.3 cm^3 6. 1475.0 cm^3 653.7 cm^2

7. 408.4 m^2 8. GH¢ 36.55   9. 2.7 cm

Review exercise 9

1(a) 520 cm^2 780 cm^3 (b) 432 cm^2 720 cm^3

(c) 342 m^2 702 m^3

2. 339.9 m^3

3(a) 81 cm^2 45 cm^3 (b) 270 cm^2 215 cm^3

(c) 569.5 cm^2 870 cm^3 (d) 31.5 m^2 11.3 m^3

4. 384   5(a) 12   (b) 15   6. 729 cm^3 7. 610.9 cm^3

8. 896 cm^2 9. 1014 cm^3 22.5 cm   10. GH¢ 13.38   11. 30 m

12. 24 m   159.7 cm^2 138.6 cm^3 13. 192 cm^2 144 cm^3

14(a) 132 cm^2 159.7 cm^2 138.6 cm^3

(b) 264 m^2 341 m^2 462 m^3

(c) 44 m^2 204.3 m^2 110 m^3

(d) 18.9 m^2 20.6 m^2 92.4 m^3

15. 1.65 m^3 16. 880 cm^2 17. 15714.3 cm^3 18. 4 cm

19(a) 204.3 cm^2 282.9 cm^2 314.3 cm^3

(b) 47.1 m^2 75.4 m^2 37.7 m^3

(c) 250.8 cm^2 404.8 cm^2 462 cm^3

(d) 219.9 m^2 333.1 m^2 377.1 m^3

20. 16 cm   21. 2 cm   22. 1.7 cm   23. 775.7 cm^2

24. 4.0 cm   25. 410.7 cm^2 10.5 cm

26. 58.9 cm^2 46.3 cm^3 27. 10.5 cm   28. 13.0 cm

29. 12.6 cm   30. 85.5^0 31. 68.7^0 64.9^0 32. 70.8^0

33. 128.1 cm^2 34. 217.6 cm^2 35. 248.4 cm^2 36. 558.6 cm^3

37. 175.9 cm^3 38. 6 m   39. 3 cm   40. 1870.9 cm^2 10677 cm^3

41. 5662.7 m^2 16,333 m^3 42. GH¢ 3575

43(a) 400.0 cm^2 999.4 cm^3 (b) 377.1 m^2 754. 3 m^3

(c) 179.1 m^2 160.3 m^3

44(a) 1257.1 cm^2 4190.5 cm^3 (b) 452.6 m^2 905.1 m^3

(c) 154 m^2 179.7 m^3

45(a) 2.3 m   (b) 3.5 cm   (c) 7.9 cm

46. 27.8 cm^2 47. 606.4 m^3 48. 1.2 cm

Chapter Test 9

1. 480 cm^2,   720 cm^3 2. 6 cm   288 cm^2 61.7 cm^3

3. 7.4 cm   4. 7128 litres   5. 23.6 cm^3 6. 53 cm

7. 28.0 cm^2 9. 2.7 cm   10. GH¢ 8,391.42

Chapter 10

Try this

1. 4198.8 km   2. 8900.9 km   3. 8937.2 km

Chapter 10.1

1(a) 2704.8 km   (b) 2188.9 km   (c) 6181.9 km   (d) 5177.7 km

2(a) 26,910.8 km   (b) 10,409.1 km (c) 20,108.8 km (d) 23,067.9 km

3. 79.2^0 4. 68.0^0 5. 56^0 6. 78.0^0

Exercise 10.2

1(a) 61^0 (b) 68^0 (c) 2^0 (d) 140^0

2. 4939.8 km   3. 3386.2 km   4. 2242.6 km   5. 1180.3 km

6. 4709.4 km   7. 72.4^0 8. 15.2^0 E   9. 39.7^0 W   10. 39^0 W

11. 549.1 km h^{-1} 12. 13.1 hr

Exercise 10.3

1(a) 36^0 (b) 93^0 (c) 64^0 (d) 30^0

2. 6702.9 km   3. 8043.5 km   4. 12,065.3 km   5. 4356.9 km

6. 5027.2 km   7. 11, 171.6 km   8. 15.6 hr   9. 547.4 km h^{-1}

10. 64.4^0 11. 76.9^0 N   12. 30.6^0 S

Review exercise 10

1(a) 41^0 (b) 70^0 (c) 20^0 (d) 113^0

2(a) 65^0 (b) 70^0 (c) 120^0 (d) 48^0

3(a) 3578.8 km   (b) 1656.4 km   (c) 5800.4 km   (d) 2905.5 km

4. 36.1^0 5. 35.1^0 W   6. 2457.7 km   7. 15,640.2 km

8. 9831.0 km   9. 5585.8 km   10. 22.4^0 11. 493.6 km h^{-1}

12. 12.3 hr   13. 42.9^0 W

14(a) 17,414.7 km   (b) 13,405.9 km

15(a) 2188.9 km   (b) 4484.7 km

Chapter Test 10

1(a) 5650.9 km   (b) 5918.3 km   2(a) 68.9^0 E   (b) 5585.8 km

3(a) 9400.8 km   (b) 783.4 km h^{-1}

www.ingramcontent.com/pod-product-compliance
Lightning Source LLC
Chambersburg PA
CBHW070219190526
45169CB00001B/15